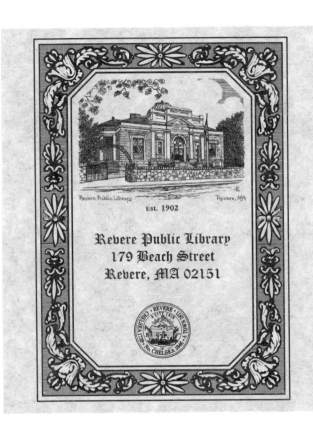

ENDANGERED
SPECIES

A Reference Handbook

Other Titles in ABC-CLIO's
CONTEMPORARY
WORLD ISSUES
Series

Books in the Contemporary World Issues series address vital issues in today's society such as terrorism, sexual harassment, homelessness, AIDS, gambling, animal rights, and air pollution. Written by professional writers, scholars, and nonacademic experts, these books are authoritative, clearly written, up-to-date, and objective. They provide a good starting point for research by high school and college students, scholars, and general readers, as well as by legislators, businesspeople, activists, and others.

Each book, carefully organized and easy to use, contains an overview of the subject; a detailed chronology; biographical sketches; facts and data and/or documents and other primary-source material; a directory of organizations and agencies; annotated lists of print and nonprint resources; a glossary; and an index.

Readers of books in the Contemporary World Issues series will find the information they need in order to better understand the social, political, environmental, and economic issues facing the world today.

ENDANGERED
SPECIES

A Reference Handbook

Clifford J. Sherry

**CONTEMPORARY
WORLD ISSUES**

ABC-CLIO

Santa Barbara, California
Denver, Colorado
Oxford, England

Cover photograph: Corel Corporation

Library of Congress Cataloging-in-Publication Data

Sherry, Clifford J.
    Endangered species : a reference handbook / Clifford J. Sherry.
       p.   cm. — (Contemporary world issues)
    Includes bibliographical references and index.
    ISBN 0-87436-810-3 (alk. paper)
    1. Endangered species. 2. Endangered species—Law and legislation. I. Title. II. Series.
QL82.S49   1998
333.95'22—dc21   98-36793

04 03 02 01 00 99 98 10 9 8 7 6 5 4 3 2 1

ABC-CLIO, Inc.
130 Cremona Drive, P.O. Box 1911
Santa Barbara, California 93116–1911

This book is printed on acid-free paper ∞ .
Manufactured in the United States of America

*To the memory of my son, Christopher Joseph Sherry,
and to my parents, Dorothy Bessie (Kohout) Sherry
and Clifford William Sherry.
Each in their own way shaped my life.*

Contents

Preface

The controversy surrounding the Endangered Species Act of 1973 and its amendments is divisive and volatile. It has led to considerable litigation, as well as significant arguing and dissent between the courts and Congress. It divides society into three unequal groups. The first is composed of the activists, from relative moderates such as the Audubon Society and the Sierra Club, who use the courts and political lobbying to achieve their goals; to the more radical, like Greenpeace, that advocate direct confrontation; to the really radical, like Earth First!, who choose to defend the natural world by direct action, civil disobedience, and eco-sabotage ("monkey-wrenching"). The second group, probably the largest, is society in general. People may or may not be aware of the controversy and may or may not have formed an opinion. The third group consists of the people and organizations that use the public lands and their resources—ranchers and the timber, mining, and petroleum/natural gas industries. The third group also includes people and organizations that are affected by infrastructure, that is, the placement of roads, bridges, dams, and so on. For example, if a new road goes from point A to point B, town X will prosper, but if it goes from point A to point C, town X will not prosper and town Y will. If the placement of the road depends on the presence of an endangered species, then the people and organizations will take an interest in the Endangered Species Act.

Is biodiversity important? If we let too many species slip into extinction, what will happen to us? How do we identify and protect endangered species? Should the needs of endangered species take precedence over the social or economic needs of humans? If so, should the humans be compensated in some way? If so, how? What will be the source of the funds?

It is our responsibility, individually and collectively, to answer these complex questions based on reason, not emotion. This book attempts to provide a balanced overview of the issues involved in the controversy. It offers the reader the tools and information needed to reach an informed opinion, based on facts, not emotion, which is needed before any action is taken.

Acknowledgments

No book is the work of a single mind. This book, which deals with a relatively controversial and volatile subject, is no exception. I have made every effort to keep my personal beliefs in the background and to provide as balanced a view of the controversy relating to the endangered species as possible.

I would like to thank my wife, Nancy C. Sherry, without whose help and encouragement this book would not have been possible.

I would like to thank my colleagues, past and present, who have acted as a sounding board for many of the ideas expressed in this book and played the role of devil's advocate. Special thanks go to Drs. William R. Klemm and G. Carroll Brown and to James H. Merritt. Wide-ranging discussions with Mr. Clyde C. Wilton have provided insights into many issues, including the ones discussed in this book.

I would also like to thank my editors at ABC-CLIO, who have been patient and understanding, as well as supportive.

I would like to thank VersusLaw (htpp://www.versuslaw. com) for providing access to the VersusLaw Opinion Library of full text decisions of the Supreme Court, the Federal Circuit Courts, and the State Appellate Courts.

Introduction 1

Thinking people everywhere would probably agree that biodiversity is important. If by action or inaction we human beings allow too many species to slip into extinction, we will soon follow—probably, as T. S. Eliot suggests, "not with a bang but a whimper."

No one knows exactly how old Earth is, but most scientists put its age at about 4.5 billion years. We do not know whether life originated on Earth or why, when, and how it first emerged. Fossilized imprints of bacteria that resemble cyanobacteria, highly evolved photosynthetic organisms that exist today, have been found in rocks 3.55 billion years old, and it is likely that life actually emerged a few hundred million years before then. During those hundreds of millions of years life has been adapting to new environments. Organisms have evolved to fill virtually all of Earth's ecological niches through natural selection, a process that derives from individual variation within species. Some individuals have characteristics that are favorable for certain environments but unfavorable for others. If an individual encounters a favorable environment, it flourishes and can pass on its genes to the next generation. If not, it is likely to be eliminated in favor of individuals whose characteristics are better suited to the particular environment.

Speciation, or the development of a new species or sub-species, is most likely to occur when environments are changing rapidly, when chromosomal mutations occur, or when a population becomes geographically separated from the rest of its species. When members of a species can no longer adapt to the demands of their environment, their numbers decrease until the species becomes extinct, to be replaced by another more efficient and adaptable species.

During its long existence, Earth has given birth to an unimaginable number and diversity of species, from jellyfish to saber-toothed tigers to bluebirds to humans. Although no one knows exactly how many species exist currently, a good estimate puts the number between 5 and 10 million, only about 1.5 million of which have been identified. Of the total number, mammals account for about 4,500, reptiles for about 5,000, birds for about 8,600, plants for between half a million and a million, and insects for over 6 million; amphibians, invertebrates other than insects, and protozoa account for the rest. The species alive today represent only about 2 percent of the species that have ever lived. The others—as many as a billion—are all extinct, most of them victims of the same process that created them, natural selection.

Extinctions, then, are a normal part of life on Earth. Most took place gradually over thousands of years, as old species were replaced by new species better fitted to the changing habitat. Others were sudden and cataclysmic, such as the famous K-T extinction, the largest in geological history. It occurred 64–66 million years ago, when, according to some scientists, a giant meteorite struck Earth and caused the extinction not only of the dinosaurs but of 85 percent of the species existing at the time.

Although most of the human-caused extinctions have occurred since the advent of the Industrial Revolution, that is, within the last 250 years, humans are also thought to have caused the series of mass extinctions that occurred worldwide in the late Pleistocene, about 50,000 years ago. In contrast to earlier mass extinctions caused by phenomena such as climate change and meteoroids, these were selective. They involved mainly the megafauna: the large herbivores and the carnivores that fed on them.

One of the best-studied examples of these early human-caused extinctions occurred about 11,000 years ago in North America, at about the time when humans first crossed the Bering Strait and began to settle the continent. Seventy species of megafauna, including mammoths, mastodons, horses, tapirs, camels, four-horned antelopes, ground sloths, peccaries, giant beavers, dire wolves, giant jaguars, and saber-tooth cats, became extinct within a relatively short time. It is likely that human hunting caused the extinction of the large herbivores, although that conclusion remains controversial. The large carnivores, probably not directly hunted, perished because they depended on these herbivores for food.

Since the Pleistocene, and particularly since the European colonization, several well-documented cases of human-caused extinctions have occurred in North America. It is estimated that 125 species of birds and mammals have become extinct here since the Pilgrims landed at Plymouth in 1620. Most of these were hunted for food, for sport, or for use in various consumer products. Others were agricultural pests, such as birds that consumed agricultural products or carnivores that developed a taste for domestic animals. In contrast, the fossil record suggests a much slower rate of extinctions in the Pleistocene: during a 3,000-year period—about ten times longer—only 50 species of mammals and 40 species of birds were lost to extinction in North America.

The bison, which survived the Pleistocene extinctions, was driven to the brink of extinction in historic times. In the mid-1800s, when the Europeans began crossing the Mississippi in significant numbers, around 65 million bison roamed the Great Plains. In the space of about 30 years, most were slaughtered for their tongues, which were eaten as delicacies; for their hides, which were made into fashionable buffalo robes; and for sport. Only about 1,000 remained by the 1880s. Today the only free-ranging bison herd left in the United States, which consists of about 1,600 individuals, lives in Yellowstone National Park. It was started from 23 individuals in 1902.

The passenger pigeon (*Ectopistes migratorius*) did not fare as well as the bison. Native to eastern North America, these birds numbered in the billions when Europeans arrived on the continent. Settlers, who told of vast flocks darkening the sky for days during migration, shot them by the wagonload for food and sport. As a result, their numbers steadily declined. Since passenger pigeons evidently required dense aggregations to breed and thrive, they could not survive when their numbers were reduced below these critical levels. The species was officially classified as extinct when the last known individual, a female named Martha, died on 1 September 1914, in the Cincinnati Zoo. A monument to the passenger pigeon in Wisconsin's Wyalusing State Park declares, "This species became extinct through the avarice and thoughtlessness of man."

The great auk, or garefowl *(Pinguinus impennis)*, was a flight-
less seabird that lived on rocky islands in the North Atlantic.
About 3 feet tall and penguinlike, the auk was a fast underwater
swimmer but awkward and defenseless on land. Enormous
numbers were killed for food and bait, particularly for their
feathers, which were used for mattress and pillow stuffing. The
last specimens were killed by two fishermen who also smashed
the last egg in June 1844 at Funk Island.

The Carolina parakeet *(Conuropsis carolinensis)*, North America's only native parrot, met the same fate at the hands of humans.
Prior to the early 1900s, these colorful birds—vibrant green and
blue with a yellow head and red eye mask—inhabited most of the
eastern United States. But their beauty, along with their choice of
food, proved their undoing. Their colorful plumage was used to
decorate women's hats, and the parakeets developed a taste for
freshly planted grain, ripe crops, and orchard fruit, which put
them at odds with farmers. The last captive Carolina parakeet,
named Incas, died in the Cincinnati Zoological Garden in 1914.

These three species, along with ten or twenty others worldwide, including the ivory-billed woodpecker, the Siberian tiger,
the African elephant, the mountain gorilla, the rhinoceros, and
the whales, have the dubious honor of being the superstars in the
modern-day extinction drama—those species that everyone has
heard about, talked about, and worried about. But they represent
only a tiny fraction of the species in danger of extinction or the
hundreds of thousands that have become extinct within our life-
time without even having been identified or named.

How Do Species Become Endangered?

Humans have always relied on other species for food, clothing,
and shelter. Often, however, we have exploited and overexploited
them because of their economic value. In North America, native
peoples lived off the wildlife population for millennia, many as
nomads—hunting in a particular area, then moving on—in a generally sustainable relationship with the ecosystem. Early European trappers used much the same system, except that they
usually hunted the animals of a region for their hides or fur until
they were so scarce that hunting was no longer profitable. Then
they moved to a new area. The decline of the beaver, the largest
rodent in North America, is one result of the business activities of
these early trappers. Beaver were avidly trapped by the first Euro-

pean trappers, since hats made from beaver fur were fashionable from the seventeenth to the early nineteenth centuries. As a result, beaver went into decline as early as 1638, and by the beginning of the nineteenth century, the species had been virtually wiped out east of the Mississippi River. When Lewis and Clark crossed the Rocky Mountains in 1805 and continued to the Pacific Coast, they sent back reports that the area was rich in beaver. But by 1840 that area, too, had been virtually stripped of beaver. Fortunately, beaver have been protected since the beginning of the century and are now doing well on both sides of the Rocky Mountains.

By introducing exotic species of animals and plants into functioning ecosystems, both accidentally and deliberately, humans have caused the extinction of some native species and altered the complex interactions of others. From about 1870 to 1890, Acclimatization Societies in a number of American cities deliberately introduced at least 18 different species of European songbirds, including a few pairs of European starlings that were released into Central Park in New York City. Today, starlings are one of the most abundant and widespread nuisance species on the North American continent, posing problems for aviation as well as agriculture. The European house sparrow was also introduced during this period. It spread rapidly across the continent, ousting many native sparrows and other birds from their accustomed habitats. The U.S. Office of Plant Introduction also claims to have introduced more than 200,000 species and varieties of plants into the United States, and many invasive plants, including the arundo weed, tumbleweed, purple loosestrife, and the kudzu vine, were introduced accidentally and now pose serious problems to both native plants and animals and to farmers and ranchers.

Most introduced species have become pests because they have no natural enemies in their new environment and can therefore outcompete native species, which do have natural enemies. Perhaps the most extreme example of this problem is the introduction of the European rabbit into Australia in 1859. Initially two dozen rabbits were brought from England and released on a ranch. Being rabbits, they did what rabbits do: reproduced and reproduced and reproduced. . . . In Europe, their natural enemies, the weasel and the fox, helped keep their numbers in check. In Australia, there were no weasels or foxes, just the dingo (another introduced species—domestic dogs—that went wild about 3,500 years ago) and Tasmanian wolves. But sheep ranchers, who worried that these predators would attack the sheep, kept them in check, and rabbits spread to all areas of

Australia except the tropical regions in the north. In 1950, Australians introduced a viral disease called myxomatosis, which killed many of the rabbits. But some survived, and did what rabbits do, reproduce, and reproduce, and reproduce. . . . And these new generations of rabbits are not susceptible to myxomatosis.

Exotic insect species are often introduced accidentally. The fire ant *(Solenopsis invicta)* is a native of South America that was accidentally introduced into the southern United States around 1940, probably at Mobile, Alabama, perhaps on shiploads of lumber. Over the past 50 years, these insects have become established throughout the South, from Texas east to Florida and north to Tennessee, with isolated pockets even farther north. Although red to reddish brown, they do not get their name from their color but from their painful sting. Workers are quite aggressive, staging swarming attacks when their nest is disturbed, and the venom from multiple stings may cause nausea, dizziness, and even death. Their mounds—up to hundreds of them per acre—have made many a farm field all but unplowable. Fire ants apparently have few or no natural enemies or pathogens in the United States. Fortunately, however, they cannot survive prolonged periods of sub-freezing temperatures and repeated freeze/thaw cycles, a factor that has kept the ants confined to the southern United States. Because of their aggressive nature, fire ants are crowding out (or killing) other insects, lizards, birds, and small mammals.

Another cause is of extinction is our species' almost insatiable need for "materials," especially wood for heating fuel, construction, and making paper. (We currently use almost 25 percent of the world's timber harvest to make paper.) Already we have used up almost 90 percent of the old-growth forests of northern California, Oregon, and Washington and more than 95 percent of the redwood forests of California. But perhaps the greatest threat to other species is humans' need for land—land for roads and other infrastructure, or for development into cities, suburbs, farms, and ranches. For example, it is estimated that tropical rain forests, home to the vast majority of the world's species, are disappearing at the rate of 1 percent per year, most of the acreage burned to make room for farm fields.

Human Impact on Natural Selection

A species can be defined as a group of organisms that interbreed to produce viable offspring. The development of such a group of closely related organisms is the result of the ongoing process of natural selection. Speciation, or the development of a new species or subspecies, is a product of natural selection. It is most likely to occur when the forces of natural selection apply the most pressure. The individuals that survive are those most perfectly suited to the conditions of their niche in the ecosystem. Over a period of time, often over hundreds or thousands of years, they diversify and evolve to become different species specializing in different ecological niches.

In the Galapagos Islands finches studied by Peter and Rosemary Grant, for example, beak length adjusted quickly (within a couple seasons) to fit the conditions. When only hard-to-crack seeds were available because of drought, the finches with longer beaks survived and most of their offspring had longer beaks. In times of floods, when selection pressure favored birds with shorter beaks that could quickly crack the abundant small seeds, the finches developed shorter beaks. But the process is incredibly complex, with other factors, such as sexual selection, playing important parts. The result of natural selection's push and pull is an astounding variety and abundance of species on Earth, each being continually reshaped to match the demands of its environment exactly.

Some species are remarkably long lived. They emerge, flourish, and remain unchanged over the millennia. Consider the horseshoe crab, which is not a crab at all but a relative of the sea scorpion, now extinct, and modern scorpions, ticks, and land spiders. Its ancestors first appeared more than 550 million years ago (Cambrian period), and the group's characteristic body structure had developed by about 400 million years ago (Devonian period). Horseshoe crabs reached their zenith between 360 and 300 million years ago (Carboniferous period), when many species evolved and flourished in a variety of habitats. Then they declined and the number of species started to dwindle—two events that usually signal extinction. But in the case of the horseshoe crabs, the decrease in the rate at which new species formed was accompanied by decrease in the rate at which living species became extinct. Thus, the species that survived the first round of extinctions tended to survive for millions of years thereafter. Today, four species of horseshoe crab survive. Three are found in the Far East, from Japan through Vietnam; the fourth, *Limulus polyphemus*, is found along the Atlantic Coast of North America, from Nova Scotia south to Mexico's Yucatan Peninsula. Its body design, with its U-shaped carapace the color of sand, is similar to that of its ancestor millions of years ago.

Or consider what is probably the oldest living tree species, *Gingko biloba*. Also called the maidenhair tree, the gingko is the only living representative of the order Ginkgoales and the family Ginkgoaceae. It dates back to the Permian period of the Paleozoic era, 245–286 million years ago. Leaves from the modern tree resemble fossilized leaves from this era. Once common in North America and Europe, it was almost destroyed during the Ice Ages in all regions except China. The gingko is an excellent example of the importance of biodiversity. First, it is a popular ornamental tree—it has a pyramidal shape and can grow to over 100 feet tall and 8 feet in diameter—because it is resistant to insects and disease. It is also resistant to pollution, so it is commonly planted in big cities. Second, and perhaps more important, its leaves contain medically useful compounds. Medicinal use of the leaves occurred as early as 2800 B.C., when the earliest Chinese materia medica suggested that they could benefit the brain. Ginkgo leaf extracts continue to be used today. In 1989, for example, more than 10 million prescriptions for ginkgo leaf extract, mostly to treat mild depression, were written in Europe.

Other species, such as the trilobite, emerge, flourish, and then diminish until they become extinct, usually for complicated reasons. Trilobites, three-segmented, three-lobed marine animals, dominated the seas when they appeared at the beginning of the Cambrian period, between 540 and 570 million years ago. Since they appeared fully developed in the Cambrian fossil record, it is likely that ancestral trilobites actually originated during the Pre-Cambrian, 3.96 billion to about 570 million years ago. More than 6,000 species of trilobites have been identified in Cambrian rocks, accounting for more than 75 percent of the fossil species. Paleontologists use trilobites to develop time correlations in Cambrian rocks. Like some modern invertebrates, trilobites had an exoskeleton that had to be shed in order for the organisms to grow. These shed exoskeletons make up a significant portion of the fossil record. They ranged in size from a few millimeters to as large as 45 cm in length. Some were apparently active predators, some were scavengers, and others ate plankton. However, despite their early success, the number of genera and species of trilobites decreased in each succeeding geological period until the Permian period, when they all disappeared in a mass extinction, leaving no modern descendants.

Major Protection Measures

The Endangered Species Act of 1973

Many measures have been instituted by governments worldwide to stem the tide of extinctions. In the United States, federal laws that protect species includes the Lacey Act (1900), the Migratory Bird Treaty Act, the Marine Mammal Protection Act, the Bald Eagle Protection Act, and most important, the Endangered Species Act of 1973 (ESA). Through the years these measures have met with occasional success. Bald eagles, for example, once considered endangered, are a common sight today along many waterways. Wolves, too, are no longer considered endangered or threatened in certain parts of their range. But even with the help of this legislation, few species' predicaments have been resolved successfully.

The federal role in protecting wildlife began with the Lacey Act of 1900, the first attempt by any government anywhere to protect wildlife. The Endangered Species Protection Act followed in 1966, but it was essentially a toothless measure. It mandated that the secretaries of the interior, agriculture, and defense protect endangered species "insofar as was practicable and consistent with the primary purposes of their departments." However, it did create the National Wildlife Refuge System and establish a policy of federal habitat acquisition and protection. The Endangered Species Conservation Act of 1969, too, was largely exhorting, although it did extend protection to invertebrate animals and allowed the Commerce Department to prohibit importation of endangered species.

In addressing the problem of extinction again in January 1973, the United States Congress observed that the country's rapid growth and technological advance, which had resulted in one of the highest standards of living in the world, had led to the extermination of some native wild animals and threatened the continued existence of other wild species. In the Endangered Species Act of 1973 "it is . . . declared to be the policy of Congress that all Federal departments and agencies shall seek to conserve endangered species and threatened species and shall utilize their authorities in furtherance of the purposes of this Act." By using that language Congress, according to the Supreme Court in *Tennessee Valley Authority v. Hiram G. Hill, et al.* (98 S.Ct. 2279), revealed a conscious decision to grant the preservation of endangered species priority over the primary missions of federal agencies and to establish the value of an endangered species as incalculable.

After nearly a year of debate, competing bills, revisions, and compromises, a conference version of a bill protecting endangered and threatened species (House Conference Report 93-740) emerged. It provided "a means whereby the ecosystems upon which endangered species and threatened species depend may be conserved, to provide a program for the conservation of such endangered species and threatened species, and to take such steps as may be appropriate to achieve the purposes of the treaties and conventions set forth." Discussions listed 109 species of wildlife—14 mammals, 50 birds, 7 reptiles—and 30 as threatened with extinction, including the black-footed ferret, the whooping crane, the eastern timber wolf, the masked bobwhite, the ivory-billed woodpecker, and the peregrine falcon. The Senate agreed to the conference version of the bill on 19 December 1973 and the House of Representatives on 20 December 1973.

When President Richard Nixon signed the bill into law (Public Law 93-205 87 Stat 884) on 28 December 1973, he stated:

> I have today signed S. 1983, The Endangered Species Act of 1973. At a time when Americans are more concerned than ever with conserving our natural resources, this legislation provides the Federal Government with needed authority to protect an irreplaceable part of our natural heritage—threatened wildlife.
>
> This important measure grants the Government both the authority to make early identification of endangered species and the means to act quickly and thoroughly to save them from extinction. It also puts into effect the Convention on International Trade in Endangered Species of Wild Fauna and Flora signed in Washington on March 3, 1973.
>
> Nothing is more priceless and more worthy of preservation than the rich array of animal life with which our country has been blessed. It is a many-faceted treasure, of value to scholars, scientists, and nature lovers alike, and it forms a vital part of the heritage we all share as Americans. I congratulate the 93rd Congress for taking this important step toward protecting a heritage which we hold in trust to countless future generations of our fellow citizens. Their

lives will be richer, and America will be more beauti-
ful in the years ahead, thanks to the measure that I
have the pleasure of signing into law today.

The Endangered Species Act of 1973 is the most strongly
worded environmental legislation passed in this country. Section 3
(6) defines an endangered species as "any species which is in danger of extinction throughout all or significant portion of its range
other than a species of the Class Insecta determined by the Secretary to constitute a pest whose protection under the provisions of
this Act would present an overwhelming and overriding risk to
man." Section 3(19) defines a threatened species as "any species
which is likely to become an endangered species within the foreseeable future throughout all or a significant portion of its range."
The act directs the secretary of the interior to "make the determination [whether a species is threatened or endangered] upon the
best scientific and commercial data available" and whether that
status is because of any of the following factors: (A) the present or
threatened destruction, modification, or curtailment of its habitat
or range; (B) overutilization for commercial, recreational, scientific,
or educational purposes; (C) disease or predation; (D) the inadequacy of existing regulatory mechanisms; (E) other natural or
manmade factors affecting its continued existence." The secretary
is then to "use all methods and procedures which are necessary to
bring any [listed] species to the point at which the measures provided pursuant to this chapter are no longer necessary."
Section 4, which defines endangered and threatened species,
has been the source of much of the litigation that has occurred as a
result of the act. Section 7, the other source of litigation, defines
and directs interagency cooperation, mandating that federal agencies must consult with the secretary of in the interior to ensure that
they "take such action necessary to insure that actions authorized,
funded, or carried out by them do not jeopardize the continued existence of such endangered species and threatened species or result in the destruction or modification of habitat of such species."

International Protection Measures

The World Conservation Union was established in Fontaine-
bleau, France, on 5 October 1948 as the International Union for
the Protection of Nature/World Conservation Union (IUCN).
The IUCN has published the Red List of Threatened Species since
the 1960s. The IUCN defines *extinct* as "there is no reasonable

doubt that the last individual has died" and *extinct in the wild* as known only to survive in cultivation, in captivity, or as a naturalized population well outside the past range." *Critically endangered* is defined as "facing an extremely high risk of extinction in the wild in the immediate future," when there has been a reduction in number of at least 80 percent over the last 10 years or three generations, the extent of occurrence is less than 100 per square kilometer or the area of occupancy is less than 10 square kilometers, and the population is less than 250 mature individuals. *Endangered* is defined as "facing a very high risk of extinction in the wild in the near future," when there is a 50 percent reduction in numbers in the last 10 years or three generations, the area of occurrence is less than 5,000 square kilometers and the area of occupancy is less than 500 square kilometers, and the population is estimated at fewer than 2,500 mature individuals. *Vulnerable* is defined as "facing a high risk of extinction in the wild in the medium near term," when there is a 20 percent reduction in numbers in the last 10 years or three generations, the area of occurrence is less than 20,000 square kilometers and the areas of occupancy less than 2,000 square kilometers, and the population is estimated at fewer than 10,000 mature individuals.

The Convention on International Trade in Endangered Species of Wild Fauna and Flora (CITES) was drawn up as an international treaty in 1973. It went into force on 1 July 1975, seeking to control overexploitation of wild species and to prevent international trade from threatening species with extinction. CITES lists threatened species that are or may be affected by trade and species that are not necessarily threatened at present but may become so unless trade in such species is subjected to strict regulation.

In order for these classifications to work, some level of human intervention is required. Someone, either a professional scientist or an amateur naturalist, must find, identify, and study a species to determine if it is threatened or endangered. This can be a formidable undertaking. The study must extend for some reasonable length of time. Most animals avoid human contact; many are nocturnal; and they are often located in inaccessible places. Many of the listed species of animals are either large animals with large home ranges or relatively small animals with small home ranges. Large animals are relatively easy to find and study, whereas small animals (such as the darters discussed below) are often discovered by accident and then can be subjected to intensive study. In contrast, intermediate-sized animals are often difficult to find and study—especially if you want to study the same animal repeatedly.

Types of Protection

Simple Protection

The Endangered Species Act (ESA) affords several means of assisting listed species, including simple protection from hunting. Sometimes, as in the case of the Mississippi or American alligator, that protection is all that is needed.

The Mississippi, or American, alligator *(Alligator mississippiensis)*, a species on the original endangered species list, was a contemporary of the dinosaur and has existed for more than 150 million years. It is found in the southeastern United States, where it lives in freshwater swamps and marshes, as well as in rivers, lakes, and small bodies of water. The naturalist William Bartram, in describing Florida's St. Johns River in 1770, said that "alligators are in such incredible numbers and so close together from shore to shore that it would have been easy to have walked across on their heads had the animals been harmless." But since that time the Mississippi alligator has been intensely hunted for its hide, particularly its belly skin, which produces high-quality leather. As a result of unregulated and unrestricted harvesting, the number of alligators gradually decreased. Although alligators were granted complete protection by a 1962 Florida law, poaching was common until the passage of the ESA. Since the 1970s, its numbers have steadily increased, and today the American alligator is considered a recovered species. Limited harvesting of alligators is now permitted in carefully controlled hunts. Alligators are also propagated and raised in captivity on alligator farms that provide more than 300,000 pounds of meat and 15,000 skins each year. In fact, alligator populations have increased sufficiently that nuisance alligators cause problems for humans, usually because of human encroachment on alligator territory. Initially, these problem alligators were relocated. But it was quickly discovered that the alligator often returned to its home range within a few days. Now, if a relocated alligator returns, it is killed. Its skin is sold and the money used to fund the alligator program.

Controlling Pollution

Sometimes pollution is the main factor that threatens a species continued existence, as in the case of the bald eagle *(Haliaeetus leucocephalus),* another species included on the original endangered species list.

The bald eagle, the national symbol of the United States since 1782, is a magnificent bird with a wing span of over 8 feet. The only eagle solely native to North America, the bald eagle is not bald. Its name came from a Middle English word *balled,* which means shining white. In the early part of the century eagle hunting was a popular sport. Furthermore, eagles were shot for their feathers and because they were thought to pose a threat to the livestock (especially sheep) and fishing industries. Their greatest enemy, however, was pollution. In 1947 bald eagles began to decline severely, and it was discovered that DDT, a common pesticide around water to control mosquitoes, was the culprit. Fish ate the DDT and the eagles ate the fish. The DDT in the tissues of the fish weakened the shells of the eagles' eggs until they simply collapsed from the weight of the mother bird. DDT was banned on 31 December 1972, and since then the numbers of eagles, which are protected by the Endangered Species Act as well as other federal and state laws, have begun to increase. Although it is still listed as endangered in the United States, estimates indicate that at least 3,000 pairs are nesting in North America.

Critical Habitats

Often it is not human hunting or poaching that threatens a species but the destruction of its habitat. If an ESA-listed species is in trouble as a result of habitat destruction, the U.S. Fish and Wildlife Service may decide to protect the species by designating a *critical habitat,* an area within which most human activities are restricted. A critical habitat can vary in size from a few acres to hundreds of square miles, but it must contain all of the things that a species needs to survive—food, water, shelter, and enough individuals to make a viable breeding population. It is important to remember, too, that a species' critical habitat is always part of a larger ecosystem whose dynamics are not totally understood.

Consider the case of the black-footed ferret *(Mustela nigripes),* an endangered species (listed 2 June 1970) that lives on the Great Plains and was thought to be extinct in the wild since 1974. These ferrets are up against a variety of obstacles. First, ferrets are *obligate associates* of prairie dogs *(Cynomys* spp.), which means that ferrets must live in association with prairie dogs to survive. The prairie dogs, named for their bark-like call, are the principal food of the ferrets. Once abundant throughout the

Great Plains, prairie dogs have been intensively hunted because they are thought to damage crops and compete with livestock for grass. But as the prairie dogs disappear, so do the ferrets. Second, ferrets not only fall prey to coyotes, they must compete with coyotes for food, as coyotes also prey on prairie dogs. In the past, the coyotes were held in check by the Mexican wolf (*Canis lupus Baileyi*, listed as endangered 11 March 1967), which hunted larger prey than prairie dogs (including coyotes) and thus did not compete with the ferrets. But the endangered wolf is no longer a threat to the coyotes. Third, the black-footed ferret is also especially susceptible to canine distemper, which is often 100 percent fatal to this species. Canine distemper is carried by coyotes and by domestic dogs.

In 1981, an isolated population of 129 ferrets was discovered in a prairie dog town near Meeteetse, Wyoming, but an outbreak of canine distemper soon severely reduced the population. The remaining ferrets were trapped and taken to a Wyoming Game and Fish research unit to set up a captive breeding program. The captive breeding program was successful, and groups of ferrets were sent to seven zoos nationwide to prevent a calamity at one location from wiping out the species. Beginning in 1991, captive bred ferrets were released back into the wild, first at the Conata Basin/Badlands Reintroduction Area in southwestern South Dakota.

Will the black-footed ferret survive? Its prognosis looks good now, but only time will tell. It is important to remember that, as the Endangered Species Act is currently written, critical habitats can be established only on public lands, that is, lands owned or controlled by federal or state governments.

The process of designating a critical habitat is described in the *Code of Federal Regulations*, Title 50—Wildlife and Fisheries, Chapter Four, Part 424—Listing of Endangered and Threatened Species and Designating Critical Habitat, Sec. 424.12, Criteria for Designating Critical Habitat. Section (b) holds that:

> In determining what areas are critical habitat, the secretary shall consider those physical and biological features that are essential to the conservation of a given species and that may require special management considerations or protection. Such requirements include, but are not limited to the following:
>
> (1) Space for individual and population growth, and for normal behavior;

(2) Food, water, air, light, minerals, or other nutritional or physiological requirements;

(3) Cover or shelter;

(4) Sites for breeding, reproduction, rearing of offspring, germination, or seed dispersal; and generally;

(5) Habitats that are protected from disturbance or are representative of the historic geographical and ecological distributions of a species.

When considering the designation of critical habitat, the secretary shall focus on the principal biological or physical constituent elements within the defined area that are essential to the conservation of the species. Known primary constituent elements shall be listed with the critical habitat description. Primary constituent elements may include, but are not limited to, the following: roost sites, nesting grounds, spawning sites, feeding sites, seasonal wetland or dry land, water quality or quantity, host species or plant pollinator, geological formation, vegetation type, tide, and specific soil types.

Recovery Plans

A recovery plan is a formal statement of what needs to be done to re-establish a species so that it can be removed from the list of threatened and endangered species. Section (f) of the Endangered Species Act describes the process of developing recovery plans. The recovery plan for the black-footed ferret, for example, requires increasing the captive population of the species to a census size of 200 breeding adults, establishing a prebreeding census population of 1,500 free-ranging breeding adults in 10 or more populations with no fewer than 30 breeding adults in any population by the year 2010, and encouraging the widest possible distribution of reintroduced populations. In order to accomplish these goals, according to the recovery plan, researchers must

• Ensure reproductive success and survival of captive *M. nigripes*

• Locate, evaluate, and maintain potential habitat (prairie dogs) in North America

• Locate additional populations of the species

• Begin continuous releases of the species into reintroduction sites when captive populations have reached or approach a sustainable population of 200

- Manage reintroduced and other populations
- Establish organizational arrangements to accomplish tasks and increase communications.

The published black-footed ferret recovery plan provides considerable detail about its goals and methods to achieve them. However, these endeavors are not inexpensive. The federal government spent about $2 million dollars on the Black Footed Ferret Recovery Program; it cost about $300,000 to set up the captive breeding program and another $200,000 to maintain it.

Captive Breeding

Captive breeding can be one of the most successful ways of dealing with a species in crisis, although it is often complicated by unforeseen requirements or difficult-to-achieve breeding conditions. Individuals are captured in the wild and brought to a zoo, university, or other facility. There they are housed under conditions believed to be right for breeding and protected from disease, predators, and human activities in the hope that they will produce offspring.

Most animal species have very specific requirements that must be met before they will breed, such as space, number of fellow species-members present, and the type and amount of shelter available. Cheetahs, for example, rarely breed in zoos because their courtship rituals require a good deal of space. One of the reasons passenger pigeons became extinct was that a successful breeding population could not be established because these pigeons require a large number of their species present before any individual pair will begin to mate. Another problem caused by a small breeding population is inbreeding and resulting genetic problems. But if enough animals are available and they can be induced to breed in captivity, then a captive breeding program can help save them from extinction—provided that funding is available. Captive breeding can be prohibitively expensive. The recovery plan for the California condor (*Gymnogypus californianus*) was drawn up in 1974. Since that time, about $20 million has been spent on trying to save this species.

The California condor is among the rarest of birds. The largest flying bird in North America, it relies on soaring rather than flapping flight. From the Pleistocene, 2 million years ago, to about 10,000 years ago, the condor's range included the coastal regions of North America from British Columbia to Baja California, east to Florida, and north to New York. Lewis and Clark reportedly sighted a condor fishing for salmon in the Columbia River and tried, but failed, to shoot it. By the early 1980s there were only 21 or 22 condors left in the wild and in captivity. Their numbers and territory had been reduced by the weakening of their eggshells caused by pesticide consumption, lead poisoning from eating lead bullets in deer and other animals killed by humans, collisions with power lines, shooting, and other human activities. In 1987 the last of the wild condors were captured to ensure their safety and to provide breeding pairs in captivity. Captive breeding programs for the condor are ongoing at the Los Angeles Zoo, the San Diego Wild Animal Park, and the World Center for Birds of Prey. The first successful captive breeding occurred in 1988. By 1995 the captive population had reached 52 individuals. On 14 January 1992 the first captive-born California condors were released at the Arundell Cliffs, and on 2 December 6 condors were released within the Sespe Condor Sanctuary. As of 27 January 1998, the captive population had risen to 93 and the wild (released) population to 39, with 15 at Vermilion Cliffs, 5 at Big Sur, and 19 at Lion Canyon/Castle Crags.

Species Preservation in Poor Countries

Many second and third world countries do not have the financial resources to address problems of human health or population growth, let alone the problems of other species. Even so, many have taken steps to protect their endangered species, usually by establishing national parks or preserves. Kenya, for example, established Tsavo East (4,536 square miles) and West (3,500 square miles) in 1948; Rwanda established Volcanoes (58 square miles) in 1929; Tanzania has set aside Gombe (20 square miles) and the Serengeti (5,000 square miles); Zambia established Kafue (8,649 square miles); and hundreds of other preserves are spread over the earth. Within the constraints of limited resources, most of these parks are well run. But they suffer from a built-in problem: while most of their revenue comes from tourism, the preservation of the wildlife often depends on leaving it undisturbed. The solution, in most cases, is to allow tourists access to portions of the park and leave large areas off limits.

Another problem related to parks is that animals do not recognize their borders and occasionally wander outside their protection. If the animal is a predator, it is usually killed because of the likelihood that it will do what predators do—kill, usually domestic animals but occasionally humans. Other wild animals, even elephants, are considered pests because they destroy crops, either by moving through them or by eating them. Often they, too, are killed.

Poaching is a problem in many parks and preserves. People kill a few animals for food. Others are captured alive and enter the illegal trade in animals, usually ending up in private zoos. A number of species are killed for their pelts, rhinos for their horn, and elephants for ivory. In fact, it is likely that the poaching of African elephants will increase with the recent decision by the Convention on International Trade in Endangered Species (CITES) to allow a resumption of the ivory trade in Botswana, Namibia, and Zimbabwe. Indian elephants are less threatened because the females are tuskless and only 60 percent of the males have tusks.

Rhinos, found in eastern and southern Africa and in tropical Asia, have existed for over 50 million years. Today they are on the verge of extinction, and the problem is their horns. The five species of rhinoceros are characterized by one (in the two species of the genus *Rhinoceros*) or two (in the other three genera) horns on the upper surface of the snout, composed not of true horn but of keratin, a fibrous protein found in hair. In Asia and the Middle East, powdered rhino horn is believed to have medicinal and restorative qualities. Some Asians believe that it is a source of youthful vigor and sexual stamina as well as a cure for a variety of diseases. In the Middle East, daggers made with a rhino horn handle are a status symbol. As a result, only about 12,000 of these magnificent animals are left in the wild. Of these, more than half are white rhinos. The remaining four species combined are represented by fewer than 5,000 individuals. (There are also about 1,000 individuals in captivity.) The rhino is protected under CITES and the U.S. Endangered Species Act. Nevertheless, poaching continues. In response, Zimbabwe instituted a program to amputate the horns of rhinos within its borders. Unfortunately, not all of the horn can be removed, and it is so valuable that poachers find it worth their while to kill the rhino to get just the stump. To make matters worse for the rhino, Zimbabwe has had to cut back on its antipoaching patrols to save money.

New Threats to Protected Wildlife under GATT

Although it was not included in the original endangered species list, Kemp's Ridley sea turtle *(Lepidochelys kempii (Garman))* is the most endangered of the sea turtles, listed as endangered throughout its range on 2 December 1970. The entire population consists of fewer than 3,000 individuals nesting on a 14.9-mile stretch of beach between Barra del Tordo and Ostional in the state of Tamaulipas, Mexico, in an area called Rancho Nuevo. The Ridleys are threatened by human activities such as collecting eggs and killing adults for meat and other products. They are also threatened by pollution caused by onshore and offshore oil exploration and production rigs, as well as the debris floating in the Gulf of Mexico, such as plastics, monofilament, and discarded netting. In addition, many Ridleys are killed in shrimp nets. The U.S. National Marine Fisheries Service has mandated the use of the turtle-excluder devices by shrimp trawlers, and efforts are being made to establish a second nesting site on Padre Island in Texas. Two thousand eggs were taken to Padre Island in an effort to imprint the hatchlings on that environment. If the plan works, the turtles will return to Padre Island, rather than the Rancho Nuevo area, when they reach breeding age. Unfortunately, there are no indications so far that the turtles are returning to Padre Island.

In April 1998, the World Trade Organization (WTO), which was established in 1995 under the General Agreements on Tariffs and Trade (GATT), ordered the United States to drop its ban on shrimp imports from nations that do not use turtle-excluder shrimp nets. The case was brought to the WTO by Thailand, Malaysia, India, and Pakistan, which argue that the U.S. regulations improperly ban the product (shrimp) based on the way it is produced. It is currently not clear if the United States will comply with the WTO ruling or what will happen if it refuses. If the United States does not comply, it could face sanctions against some of its exports.

Controversies Surrounding the ESA

The Problem of Defining Species

Foremost among the questions that the ESA raises is this: What exactly is a species? The language of the act suggests that taxonomists and systematists (those who specialize in classifying living organisms) and the vast array of "ologists"—zoologists, mammalogists, herpetologists, ichthyologists, entomologists, and so on—agree on a definition of species. Most scientists would, in fact, agree with this general definition: A species is a group of successful organisms that shares an environmental niche, tend to have similar morphological and behavioral characteristics, and can mate and produce viable offspring with one another. Individuals of a single species share a common gene pool that embodies the genetic continuity of life.

As with most generalities, however, the devil is in the details. Consider a simple example. It is well established that the Saint Bernard and the Chihuahua are members of the same species, *Canis familiaris,* and virtually every scientist would agree with that classification. Most would also agree that the ability of two organisms to mate and produce viable young is the essential element in the definition of a species. Do the Chihuahua and St. Bernard meet this essential criterion? Hardly. And what if scientists unfamiliar with this species came across these two animals in the field? Would they classify them as a single species? Probably not.

These are the kinds of conundrums that confront scientists when they try to apply the concept of species, a human abstraction, to the real world. Since plants and animals do not come equipped with stickers that identify them, scientists have developed a standardized procedure for establishing a new species: they describe and measure, in great detail, the first male and female specimens (known as *type specimens*); they determine what taxonomic group they belong to—what order, family, genus; then they name the new species according to formal scientific convention, using two or more Latin words, the first of which indicates genus, the second species. These two words together make up the species name. The physical description is then applied to any additional specimens to determine whether they are examples of this new species or a species that has already been described.

Mistakes are common. Many species initially described as new were later found to be only forms or variants of species that had already been described, and the relationship between very closely related species is often hotly debated. It is also important to remember that the vast majority of species (especially insects) have not yet been identified, and even among organisms that have been identified, described, and classified, many are known from a single specimen or a few specimens in some museum. Thus it is often difficult even to make an accurate estimate of the number of individuals that make up the species. The costly, time-consuming process of determining whether its numbers are increasing or decreasing often requires decades or more.

Endangered Wildlife versus Property Rights and Jobs

Since the Endangered Species Act of 1973 is the strongest wildlife protection measure ever passed by the U.S. Congress, it is not surprising that it has been controversial. The authorization of the Endangered Species Act expired on 30 September 1992. In 1995 the House of Representatives agreed by voice vote to stop all listings under the act until it is reauthorized, and the Senate prohibited the listing of additional species until the end of 1995 or the reauthorization of the act. The listing moratorium was extended until 1996 or reauthorization of the act; on 26 April 1996 President Clinton lifted the moratorium. Congress granted him this power because of his threat to veto the Fiscal Year (FY) 1996 funding bill for the Department of the Interior and other segments of the government. Implementation of the act has continued since then via annual appropriations for the Departments of Interior and Commerce. The battle over its reauthorization, still ongoing in 1998, has been bloody, pitting ranchers, timber and mining companies, and developers against environmentalists. It is unlikely that the law will emerge from the process intact. Critics' objections arise primarily from the restrictions involved in critical habitat preservation and property rights, as well as a legal issue known as *taking*.

When the Endangered Species Act was fully authorized and in full force, it applied only to activities on public lands or activities supported by federal dollars, based on the theory that public lands are owned by all citizens in common and activities supported by federal dollars are for the common good. In practice, however, public lands are often used for private purposes. Ranchers graze sheep or cattle on public lands, paying a fee for this privilege. Timber harvests on public lands, for example, are federally subsidized, and minerals are extracted from leased federal land. Many ranchers, timber companies, and mining operations have come to rely on these long-standing arrangements. If the privilege were withdrawn or significantly altered by the restrictions on critical habitat for endangered species, job loss and business failure could and probably would follow. Likewise, infrastructure projects supported by federal funds, such as roads and dams, have significant impacts on the social and economic development of the affected areas. In many cases concern for biodiversity seems to be pitted against the economic survival of individual families or entire towns.

However, the Endangered Species Act and its amendments mandate that human social or economic considerations not be allowed to affect the determination of whether a species is threat-ened or endangered. Nor are these considerations allowed to af-fect the designation of critical habitat. If the Endangered Species Act is reauthorized, it is likely that it will be modified, perhaps significantly (see Chapters 4 and 5). A change supported by one group would include considerations of the social and economic impact of attempting to protect specific species and designations of critical habitat. Another group would support extending the authority of the act beyond public lands and federal projects to allow designation of critical habitat to apply to private lands.

The Fifth Amendment of the Constitution states that "private property shall not be taken for public use without just compensa-tion." Interpretation of this amendment by the Supreme Court has protected individuals (or corporations) from the "physical" taking of private property (for example, land, money, goods) by the gov-ernment without compensation. The degree to which the amend-ment protects property owners from partial or regulatory taking (a reduction in the value of a property caused by regulatory activities) remains controversial. If, for example, a federal designation of wet-land prevented owners from building a house on property they had bought for that purpose, a takings question arises. Should the own-ers be compensated for the decrease in the value of their property, even though no physical taking has occurred? Environmentalists generally say no—that the owner has no right to build a house that is guaranteed by the Constitution and that compensation in such cases will make enforcement of the Endangered Species Act and other wildlife protection laws prohibitively expensive. Theoreti-cally, the Supreme Court will attempt to balance everyone's rights. However, in the two important cases decided so far, it has awarded compensation to owners. If regulatory activity under the act is ex-tended to private property and the Court holds that this is a regu-latory taking, then private landowners would most likely have to be compensated either by direct payment or by a tax deduction.

Mixed Messages from Congress and the Courts: Two Case Studies

The Darters

Darters are small fish that belong to the perch family (Percidae), which includes perch, walleye, and sauger. Approximately 150 species live in North American waters. In the United States, they are distributed in northern temperate zones. Of the 102 fish listed as threatened or endangered as of 1994, 14 were various species of darters. Two in particular, the snail darter and the fountain darter, have been the focus of Congress, the public, and the courts.

In August 1973, David A. Etnier, a University of Tennessee ichthyologist, discovered a previously unknown species of perch in the Little Tennessee River. It is about 3 inches long, with a dorsal fin divided into two separate fins, brown to brownish gray with a trace of green. It lives for one to two years and eats aquatic gastropods (snails). Etnier described this fish and named it *Percina (Imostoma) tanasi*, the snail darter.

About the same time, the Tellico Dam, a project designed to provide electricity, flat-water recreational opportunities, and flood control was nearing completion. When it impounded the river, it would destroy all of the known habitat of the snail darter. Thus in January 1975, several individuals petitioned the secretary of the interior to list the snail darter as an endangered species and prevent completion of the Tellico Dam. The snail darter was listed on 8 October 1975. In making the listing the secretary noted that "the snail darter is a living entity, which is genetically distinct and reproductively isolated from other fishes." Its critical habitat, which was described in 41 Fed.Reg. 13926-13928 (see also 50 CFR 17.81), would be destroyed by the operation of the dam. On 28 February 1976, suit was brought in district court to permanently prevent completion of the dam (see 419 F.Supp. 753, Chapter 5). The court found that completing the dam would probably jeopardize the snail darter's continued existence, but also that Congress, aware of the snail darter problem, had continued to fund the project. The court concluded that "at some point in time a federal project becomes so near completion and so incapable of modification" that a court where justice is administered according to fairness, rather than with strictly formulated rules (that is, an equity court) should not apply a statute enacted long after inception of the project." However, the court of appeals (549 F.2d 1064, see Chapter 5) reversed the district court's decision and ordered the project halted "until Congress, by appropriate legislation, exempts Tellico from compliance with the Act or the snail darter has been deleted from the list of endangered species or its critical habitat materially redefined."

On appeal, the Supreme Court (98 S.Ct. 2279, see Chapter 5) held that completion and operation of the dam would either eradicate a known population of the snail darter, an endangered species, or destroy its critical habitat. Therefore, it prohibited completion of the dam, even although the dam was almost finished and Congress had continued to appropriate large sums of public money for the project. The Court held that Section 7 of the ESA was not limited to projects in the planning stages and that Congress intended to halt and reverse the trend toward species extinction, whatever the cost. The language of the ESA revealed a conscious decision to give endangered species priority over the primary missions of federal agencies and clearly showed that Congress viewed the value of an endangered species as incalculable. Continued appropriations did not constitute implied repeal of the Endangered Species Act as applied to the project. Further, the Court reiterated that its "appraisal of the wisdom or unwisdom of particular course consciously selected by Congress is to be put aside in process of interpreting a statute. Once meaning of enactment is discerned and its constitutionality determined, judicial process comes to an end. We do not sit as a committee of review, nor is a court vested with power of veto." However, on 25 September 1979 President Jimmy Carter refused to veto a congressional bill directing that the Tellico Dam be completed and ESA restrictions waived.

The fountain darter (*Etheostoma fonticola*) is a reddish-brown fish less than 22 mm long that eats various invertebrates. It is found only in the Comal and San Marcos springs and in the San Marcos and Comal rivers, part of the Guadalupe River basin in southern Texas. The springs, which discharge into the river, represent the outflow of the Edwards (Balcones Fault Zone) Aquifer, a unique carbonate aquifer located in a porous, honeycombed formation composed of Edwards and associated limestones. The factors that affect the fountain darter are its extremely limited range and the flow of water in the springs and rivers where they feed. The population is estimated at approximately 100, 000.

The fountain darter was listed as endangered on 13 October 1970, and since its listing has been the subject of several lawsuits (see 995 F.2d 571; 1997.CO5.174; 1997.CO5. 256 (http://www.versuslaw.com). The Sierra Club, which is the plaintiff in each case, is suing either Bruce Babbitt, as the secretary of the interior, or the city of San Antonio and others who withdraw water from the aquifer. Each of these suits was aimed at establishing some control over the amount of water that is removed from the aquifer in order to keep water flowing in the springs and the Guadalupe River, which is the habitat of not only the fountain darter, but four other listed species: the Texas blind salamander (*Typhlomolge rathbuni*); the San Marcos gambusia (*Gambusia georgei*), which may already be extinct; the San Marcos salamander (*Eurycea nana*); and Texas wild rice (*Zizania texana*). Partly as a result of these suits and in an effort to maintain local control over the aquifer, the Texas legislature enacted the Edwards Aquifer Act, which established the Edwards Aquifer Authority. It will control the amount of water withdrawal from the aquifer.

But difficulties arose from the fact that the Edwards Aquifer is not only the habitat of several endangered species, but also the sole source of drinking water for over a million people in San Antonio and the San Antonio–Austin corridor. Here is a clear example of the needs and desires of a large group of human beings coming into direct conflict with the continued existence of endangered species. When the next drought occurs—and it is only a matter of time—the Edwards Aquifer Authority will face very difficult legal, environmental, and moral decisions.

The Northern Spotted Owl

Four centuries ago, the landscape that greeted the first European settlers to arrive in North America was heavily forested. Today, only about 10 percent of the land that was once forested remains so. The forests of New England, the South, and the Midwest have all but disappeared. The only significant remnant of the original forest is located on the Pacific slope, extending from the Alaska Panhandle to San Francisco Bay. These virgin forests (often called *old-growth forests*) are complex ecosystems, valuable from many perspectives. They contain some of the oldest and largest trees in the world, up to 300 feet tall and 50 feet in circumference, so it is not surprising that timber companies are eager to log them. However, intact old-growth forests provide environmental services that, in the long run, are far more valuable than lumber. They help regulate the climate (both locally and at some distance), maintain water levels, prevent flooding, clean the air by storing carbon dioxide, and enrich the soil.

The old-growth forest is also home to at least six species protected by the Endangered Species Act, including the northern spotted owl (*Strix occidentalis lucida*), a species that has become the center of bitter controversy. This famous owl is medium-sized (males weigh about 1.5 pounds and are 16 to 19 inches long) and chocolate brown with light spots on the top of the head and the back of the neck. Northern spotted owls generally begin reproducing when they are about three years old and typically produce owlets every other year. Juvenile survivorship is remarkably low, however; it is estimated that 80 percent of the young die before or during dispersal from the nest.

The northern spotted owl is of particular interest to scientists and environmentalists, because it is considered an *indicator species*, which means that its success is a sort of barometer that indicates the health of the entire ecosystem. The owl eats squirrels, mice, and other small animals that eat the seeds and nuts of the forest trees and the mycorrhizzal fungus that infect the roots of the Douglas fir and other trees. If nuts, seeds, and fungi are plentiful, then there should be plenty of small animals—and spotted owls. If food is not plentiful, then the number of owls declines. In short, if the owl survives in healthy numbers, the forest is healthy. Spotted owl habitat is located in about 3 million acres currently managed by the U.S. Forest Service and the Bureau of Land Management.

Calls to protect the spotted owl and its habitat began in earnest in 1974. Researchers asked that 120,000 acres be set aside to maintain approximately 400 nesting pairs of owls. This request was rejected by the Forest Service and the Bureau of Land Management. Through 1904, the Forest Service produced a number of draft environmental impact statements and ultimately involved the National Park Service, the Fish and Wildlife Service, and the Bureau of Land Management in an interagency committee (the Thomas Committee), which produced a series of recommendations in May 1980. After a good deal of controversy and litigation (see, in particular 716 F. Supp. 479), the spotted owl was listed as threatened with extinction on 26 June 1990.

In response to ongoing litigation, Congress enacted section 318 of the Department of Interior and the Related Agencies Appropriations Act of 1989 (Hatfield-Adams Appropriations Bill). This law, commonly known as the Northwest Timber Compromise, stated that the Forest Service's current management proposal for federal old-growth timber sales, the subject of ongoing litigation, constituted "adequate consideration" for meeting the statutory requirements of the Endangered Species Act. However, the new law itself soon became the subject of litigation. The Ninth Circuit Court (see 914 F.2d 1311) found that in passing an act addressing ongoing litigation, Congress had exceeded its constitutional authority. The appeals court found that the act invaded the province of the judicial branch in violation of the separation of powers doctrine. The Supreme Court (see 112 S.Ct. 1407) reversed the Ninth Circuit Court's holding, stating that the

Northwest Timber Compromise compelled changes in law, not results under old law. It replaced the legal standards underlying the pending litigation with those set forth in the compromise's sections.

In the meantime, as the lawsuits made their way through the courts, animosity between timber interests, including local logging communities, and environmental interest groups heated to the boiling point, erupting into threats and demonstrations. In the spring of 1993, President Bill Clinton held a "forest summit" to try to end the timber problems in the Pacific Northwest. Both environmentalists and the timber industry were invited to present their views. In response to their concerns, the president then proposed cutting two-thirds less timber than the amount cut in the 1980s, but neither the environmentalists nor the timber industry appears happy with the proposal. The controversy continues in both the courts and Congress.

Conclusion

Besides the issues discussed above, the Endangered Species Act raises compelling practical and philosophical questions. What do we do when we do not know the biological needs of a potentially threatened or endangered species? What do we do when the habitat an endangered species requires no longer exists in the species' historic range? What do we do when an endangered predator is preying on an endangered prey? When funds and manpower are limited, as they almost always are, who sets the priorities among the various species? Who decides when a species is extinct? What criteria do they use? When the best evidence indicates that a species will become extinct regardless of human efforts, what action, if any, should be taken? And potentially the most compelling question of all: If we mandate that we will maintain the conditions that existed in an ecosystem during a particular time in order to protect a threatened or endangered species, are we preventing the process of natural selection from occurring? In other words, if we "artificially" protect a given species and prevent it from going extinct, do we thereby prevent the emergence of a more efficient and adaptable species that might have filled the space in the ecological niche left by the species that became extinct?

These are only a few of the difficult questions that must be answered during the debate over the reauthorization of the Endangered Species Act. The outcome of the political battle is impossible to foretell. As long as humans continue to allow the extinction of species at the current alarming rate, the outcome of the ongoing battle between humans and other species is, sadly, much easier to predict.

Endangered and Threatened Species Lists

As of 31 May 1997 there were 635 species of plants and 447 species of animals listed, while 91 species of plants and 28 species of animals have been proposed for listing. Of these, critical habitat has been designated for 124 species and proposed for 7. There are 76 species of plants and 97 species of animals that are U.S. Fish and Wildlife Service candidate species. Recovery plans have been approved for 653 species, while recovery plans are under development for 398 species. Eleven species of animals and 33 plants were added between 31 May 1997 and 31 January 1998. Of these, 343 animals and 553 plants were listed as endangered and 115 animals and 115 plants were listed as threatened. There are 521 foreign animal and 1 foreign plant species listed as endangered, while 38 foreign animal and 2 foreign plant species are listed as endangered.

This information is periodically updated on the World Wide Web page of the Fish and Wildlife Service (http://www.fws.gov). The listed endangered (E) and threatened (T) mammals include:

E   Bat, gray (*Myotis grisescens*)
E   Bat, Hawaiian hoary (*Lasiurus cinereus semotus*)
E   Bat, Indiana (*Myotis sodalis*)
E   Bat, lesser (=Sanborn's) long-nosed (*Leptonycteris curasoae yerbabuenae*)
E   Bat, little Mariana fruit (*Pteropus tokudae*)
E   Bat, Mariana fruit (*Pteropus mariannus mariannus*)
E   Bat, Mexican long-nosed (*Leptonycteris nivalis*)
E   Bat, Ozark big-eared (*Corynorhinus* [=*Plecotus*] *townsendii ingens*)
E   Bat, Virginia big-eared (*Corynorhinus* [=*Plecotus*] *townsendii virginianus*)
T   Bear, American black (*Ursus americanus*)
T   Bear, grizzly (*Ursus arctos*)
T   Bear, Louisiana black (*Ursus americanus luteolus*)
E   Caribou, woodland (*Rangifer tarandus caribou*)
E   Cougar, eastern (*Felis concolor couguar*)

E Deer, Columbian white-tailed (*Odocoileus virginianus leucurus*)
E Deer, key (*Odocoileus virginianus clavium*)
E Ferret, black-footed (*Mustela nigripes*)
E Fox, San Joaquin kit (*Vulpes macrotis mutica*)
E Jaguar (*Panthera onca*)
E Jaguarundi (*Felis yagouaroundi cacomitli*)
E Jaguarundi (*Felis yagouaroundi tolteca*)
E Kangaroo rat, Fresno (*Dipodomys nitratoides exilis*)
E Kangaroo rat, giant (*Dipodomys ingens*)
E Kangaroo rat, Morro Bay (*Dipodomys heermanni morroensis*)
E Kangaroo rat, San Bernardino Merriam's (*Dipodomys merriami parvus*)
E Kangaroo rat, Stephens' (*Dipodomys stephensi* [including *D. cascus*])
E Kangaroo rat, Tipton (*Dipodomys nitratoides nitratoides*)
T Lion, mountain (*Felis concolor* [all subspecies except *coryi*])
E Manatee, West Indian (=Florida) (*Trichechus manatus*)
E Mountain beaver, Point Arena (*Aplodontia rufa nigra*)
E Mouse, Alabama beach (*Peromyscus polionotus ammobates*)
E Mouse, Anastasia Island beach (*Peromyscus polionotus phasma*)
E Mouse, Choctawhatchee beach (*Peromyscus polionotus allophrys*)
E Mouse, Key Largo cotton (*Peromyscus gossypinus allapaticola*)
E Mouse, Pacific pocket (*Perognathus longimembris pacificus*)
E Mouse, Perdido Key beach (*Peromyscus polionotus trissyllepsis*)
E Mouse, salt marsh harvest (*Reithrodontomys raviventris*)
T Mouse, southeastern beach (*Peromyscus polionotus niveiventris*)
E Ocelot (*Felis pardalis*)
T Otter, southern sea (*Enhydra lutris nereis*)
E Panther, Florida (*Felis concolor coryi*)
T Prairie dog, Utah (*Cynomys parvidens*)
E Pronghorn, Sonoran (*Antilocapra americana sonoriensis*)
E Rabbit, Lower Keys (*Sylvilagus palustris hefneri*)
E Rice rat (=silver rice rat) (*Oryzomys palustris natator*)
E Sea-lion, Steller (=northern), western population (*Eumetopias jubatus*)
T Sea-lion, Steller (=northern), eastern population (*Eumetopias jubatus*)
E Seal, Caribbean monk (*Monachus tropicalis*)
T Seal, guadalupe fur (*Arctocephalus townsendi*)
E Seal, Hawaiian monk (*Monachus schauinslandi*)
T Shrew, Dismal Swamp southeastern (*Sorex longirostris fisheri*)
E Squirrel, Carolina northern flying (*Glaucomys sabrinus coloratus*)
E Squirrel, Delmarva Peninsula fox (*Sciurus niger cinereus*)

E   Squirrel, Mount Graham red (*Tamiasciurus hudsonicus grahamensis*)
E   Squirrel, Virginia northern flying (*Glaucomys sabrinus fuscus*)
E   Vole, Amargosa (*Microtus californicus scirpensis*)
E   Vole, Florida salt marsh (*Microtus pennsylvanicus dukecampbelli*)
E   Vole, Hualapai Mexican (*Microtus mexicanus hualpaiensis*)
E   Whale, blue (*Balaenoptera musculus*)
E   Whale, bowhead (*Balaena mysticetus*)
E   Whale, finback (*Balaenoptera physalus*)
E   Whale, humpback
E   Whale, right (*Balaena glacialis* [including *australis*])
E   Whale, Sei (*Balaenoptera borealis*)
E   Whale, sperm (*Physeter macrocephalus* [=*catodon*])
E   Wolf, gray (*Canis lupus*)
T   Wolf, gray (*Canis lupus*)
E   Wolf, red (*Canis rufus*)
E   Woodrat, Key Largo (*Neotoma floridana smalli*)

As described above, the International Union for the Protection of Nature/World Conservation Union (IUCN) and the Convention on International Trade in Endangered Species of Wild Fauna and Flora (CITES) are two international organizations that are chartered to monitor Earth's fauna and flora to determine which are in danger of becoming extinct and developing means and methods to prevent these species from becoming extinct. The IUCN lists 169 mammals, 168 birds, 41 reptiles, and 18 amphibians as critically endangered, 315 mammals, 235 birds, 59 reptiles, and 31 amphibians as endangered, and 612 mammals, 704 birds, 153 reptiles, and 75 amphibians as vulnerable. Details of other taxonomic groups as well as the actual lists of species can be found on IUCN's World Wide Web page (http://w3.iprolink.ch/iucnlib/info_and_news/index.html).

CITES lists 254 species, subspecies, and populations of mammals, 160 birds, 71 reptiles, 14 amphibians, 8 fish, 69 invertebrates, and 314 plants as species threatened with extinction which may or may not be affected by trade. More details, as well as the actual lists of species can be found at the World Wide Web page of CITES (http://www.wcmc.org.uk/CITES/).

The Conservation International (World Wide Web page http://www.conservation.org/web/fieldact/hotspots/hot97.htm) has identified 19 areas as global biodiversity hotspots, as shown below.

Biodiversity Areas	Total Endemic Vascular Plants	Total Endemic Vertebrate Fauna
The Tropical Andes (Venezuela, Colombia, Ecuador, Peru, and Bolivia)	20,000	878
Madagascar	8,000	630
Brazil's Atlantic Forest Region	6,000	532
The Philippines	6,000	462
Meso-American forests	9,000	635
Wallacea (eastern Indonesia)	3,000	455
Western Sunda (in Indonesia, Malaysia, and Brunei)	6,750	291
South Africa's Cape	7,550	92
The Antilles	7,000	347
Brazil's Cerrado	4,200	48
The Darién and Chocó of Panama, Colombia, and Western Ecuador	3,760	405
Polynesia and Micronesian Island Complex, including Hawaii	3,275	117
Southwestern Australia	2,830	71
The Eastern Mediterranean region	2,770	15
The Western Ghats of India and the island of Sri Lanka	2,687	310
The Guinean forests of West Africa	2,960	171
Eastern Sundaic Region	3,000	455
New Caledonia	2,551	81

References

Blumm, M. C. (1991) "Ancient Forest, Spotted Owls, and Modern Public Land Law." *Boston College Environmental Affairs Law Review* 18: 605–622

Bonnett, M,. and Zimmerman, K. (1991) "Politics and Preservation: The Endangered Species Act and the Northern Spotted Owl." *Ecology Law Quarterly* 18: 105–171.

Foley, E. A. (1992) "The Tarnishing of an Environmental Jewel: The Endangered Species Act and the Northern Spotted Owl." *Journal of Land Use and Environmental Law* 8: 253–283

Grierson, K. W. (1992) "The Concept of Species and the Endangered Species Act." *Virginia Environmental Law Journal* 11: 463–493.

Goplerud, C. P. (1979) "The Endangered Species Act: Will It Jeopardize the Continued Existence of Species?"*Arizona State Law Journal* 487–510.

Hare, T. (ed.) (1994) *Habitats.* New York: Macmillan.

Hartmann, W. K., and Miller, R. (1991) *The History of Earth: An Illustrated Chronicle of an Evolving Planet.* New York: Workman.

Mayr, E. (1963) *Animal Species and Evolution.* Cambridge, MA: Harvard University Press.

Raup, D. M. (1991) *Extinction: Bad Genes or Bad Luck?* New York: W. W. Norton.

Rohlf, D. J. (1989) *The Endangered Species Act: A Guide to Its Protections and Implementation.* Stanford, CA: Stanford Environmental Law Society.

Rosenberg, R. H. (1980) "Federal Protection of Unique Environmental Interests: Endangered and Threatened Species." *North Carolina Law Review* 58: 491–559.

Sher, V. M., and Hunting, C. S. (1991) "Eroding the Landscape, Eroding the Laws: Congressional Exemptions from Judicial Review of Environmental Laws." *Harvard Environmental Law Review* 15: 435–491.

Silverstein, A., Silverstein, V., and Silverstein, R. (1994) *The Spotted Owl.* Brookfield, CT: Millbrook Press.

Simmons, R. M. (1978) "The Endangered Species Act of 1973." *South Dakota Law Review* 23: 302–325.

Ward, P. D. (1992) *On Methuselah's Trail: Living Fossils and the Great Extinctions.* New York: W. W. Freeman.

Weiner, J. (1994) *The Beak of the Finch: A Story of Evolution in Our Time.* New York: Alfred A. Knopf.

Chronology

1735   Carolus Linnaeus (Carl Von Linne) publishes *Systema naturae*, introducing the Latin binomial system of nomenclature, in which the first name is the genus and the second the species.

1737   Linnaeus publishes *Genera planto rum*, in which he explains his method of systemic botany.

1749   Comte de Buffon provides the modern definition of species (a group of organisms capable of breeding and producing viable and fertile offspring).

1785   Land Ordinance Act mandates that all public domain lands be surveyed and laid out in regular plots 1 mile square (640 acres).

1827   John James Audubon begins the publication of *Birds of America.*

1831   Charles Robert Darwin joins the crew of H.M.S. *Beagle* as the ship's naturalist.

1833   Jean-Louis-Rudolphe Agassiz publishes the first volume of his studies of fossil fishes.

        The last known quagga, a relative of the zebra, dies in an Amsterdam zoo.

1835   Darwin visits the Galapagos Islands and closely observes finches (called Darwin's finches today) that seemed to have developed from a common ancestor on the mainland of South America.

1859   Darwin publishes *On the Origin of Species by Means of Natural Selection.*

1862   The Homestead Act allows settlers to obtain clear title to a quarter section of public land by living on it for five years and improving it.

1864   George Perkins Marsh publishes *Man and Nature,* which explains his findings that removing woods and vegetative cover resulted in floods, soil erosion, and fluctuations in stream flow.

1865   Gregor Mendel publishes his theories of genetics.

1866   Ernst Haeckel coins the term *ecology* to describe how organisms interact with one another in a shared habitat.

1871   The U.S. Fish and Fisheries Commission is authorized by Congress, with Spencer F. Baird as first commissioner.

        Darwin publishes *The Descent of Man and Selection in Relation to Sex.*

1872   Congress sets aside 2 million acres of public domain land to create Yellowstone National Park to serve as a "pleasuring ground for the people" in perpetuity.

1884   Congress passes the Lacey Yellowstone Protection Act.

1886   George Bird Grinnell proposes an organization to protect the nation's birds; the organization became the Audubon Society.

1887   George Bird Grinnell and Theodore Roosevelt form the Boone and Crockett Club to help stop the slaughter of big game animals.

1891   Benjamin Harrison signs the Forest Reserve Act, which created the National Forest System. Harrison quickly withdraws 13 million acres from the public domain and creates 15 forest reserves.

1900   Congress passes the Lacey Act to preserve and restore wild birds.

1901   Theodore Roosevelt becomes president after the assassination of William McKinley.

1903   John Muir, with Henry Senger, William D. Armes, Joseph Le Conte, and Warren Olney, forms the Sierra Club.

1905   The Bureau of Biological Survey is established (later combined with the Bureau of Fisheries to become the Fish and Wildlife Service).

1906   Theodore Roosevelt signs the bill that makes Yosemite the second national park.

               Antiquities Act allows the creation of national monuments.

1909   Wilhelm Johannsen coins the term *gene* to describe the carrier of heredity.

1916   The National Park System is created.

1933   The last known Tasmanian wolf dies in a zoo.

1936   The Convention Relative to the Preservation of Fauna and Flora in Their Natural State is proclaimed.

1937   The Convention between the United States of America and Mexico for the Protection of Migratory Birds and Game Mammals is proclaimed.

        Genetics and the Origin of Species is published by Theodosius Dobzhansky.

1942   The Convention on Nature Protection and Wildlife Preservation in the Western Hemisphere is proclaimed.

1945   The International Convention for the Regulation of Whaling is proclaimed.

1948   The International Union for the Protection of Nature/ World Conservation Union is established.

1957   G. E. Hutchinson defines the concept of ecological niche.

1960   National forests are directed to be managed under the principles of multiple use and sustained yield.

1962   *Silent Spring* is published by Rachel Carson.

1963   The International Convention for the Protection of Birds is proclaimed.

1964   The Wilderness Act is passed, giving legislative status to certain existing wilderness areas and providing the machinery to give legislative status to others.

        The Federal Land and Conservation Fund is established to provide funds for the acquisition of recreational facilities by federal, state, and local agencies.

1966   Congress passes the Fish and Wildlife Conservation Act, which directs the secretaries of interior, commerce, and defense to preserve habitats of species "where practicable and consistent with their program purposes."

        Konrad Lorenz , in *On Aggression*, claims that humans

are the only species to intentionally kill its own members.

1969   The National Environmental Policy Act is signed into law. It declares as a national policy that the United States will encourage productive and enjoyable harmony between humankind and the environment.

1970   April 22, 1970, the first Earth Day, starts as a student-led campus movement; initially observed on March 21, Earth Day has become a major educational and media event to sum up current environmental problems of the planet.

1973   President Nixon signs the Endangered Species Act of 1973 (ESA) into law on 28 December.

1975   The Convention of International Trade in Endangered Species of World Fauna and Flora (CITES) is signed.

1976   Endangered Species Act Amendments are passed.

1978   President Jimmy Carter signs the Endangered Species Act (ESA) of 1978 (Public Law [PL] 95-362) into law on 10 November.

1979   James Lovelock developed the GAIA theory, which suggests that the earth is a single, living organic being.

1980   The U.S. Supreme Court rules that a microbe developed by General Electric can be patented.

1982   President Reagan signs the Endangered Species Act Amendments of 1982 (PL 97-304) into law on 13 October.

1986   All known black-footed ferrets in the wild are captured and placed in a breeding program.

1988   President Reagan signs the Endangered Species Act Amendments of 1987 (PL 100-478) into law on 7 October.

1988 *(cont.)*	The U.S. Patent Office issues the first patent for a vertebrate.

The African Elephant Conservation Act is passed.

1989   Congress passes the Northwest Timber Compromise.

1992   Authorization of the Endangered Species Act expires on 30 September, allowing implementation under the act to continue by annual appropriations for the Departments of Commerce and Interior.

1995   House Resources Committee Chairman Don Young announces the formation of a House Endangered Species Task Force, chaired by Congressman Richard Pombo.

On 23 February, the House of Representatives agreed to halt all listing of species under ESA until it is reauthorized.

On 16 March, on an amendment to the Defense Supplemental Appropriation bill, the Senate agreed to prohibit the listing of additional species and further designation of critical habitat under ESA until the end of fiscal year (FY) 1995 or the act is reauthorized.

On 10 April President Clinton signs H.R. 889 into law, codifying the ESA moratorium.

1996   On 18 July the House passes the FY 1996 funding bill for the Department of Interior, which eliminates listing and prelisting activities under the ESA until the act is reauthorized and eliminates the National Biological Service as an independent agency.

On 9 August the Senate passes the FY 1996 funding bill, which extends the moratorium on additional listings until September 1996 or until reauthorization.

On 16 April President Clinton lifts the moratorium on listings.

In July, the National Governors' Association adopts a resolution calling for the reform of the ESA.

On 13 November, the Supreme Court, in *Bennett v. Spear*, hears a case that would allow people who have suffered economic harm as a result of efforts to protect endangered species to sue the secretary of the interior.

1997   On 31 January, Senators Kempthorne and Chafee released a discussion draft of their proposed ESA reform legislation.

On 19 March, in *Bennett v. Spear*, the Supreme Court reverses lower court decisions that held that only parties seeking more stringent protections of threatened or endangered species had standing to sue under the ESA "citizen suit" provision and finds that federal courts must allow economically impacted parties an opportunity to sue.

Biographical Sketches 3

Edward Abbey (1927–1989)

Edward Abbey was born in Indiana, Pennsylvania, on 29 January 1927. He left the family farm to hitchhike and ride the rails across the Midwest to the West Coast. Returning via the Southwest, he fell in love with the desert. He received graduate and postgraduate degrees from the University of New Mexico. For 15 years he worked as a part-time ranger and fire lookout at several different national parks, including two seasons as a ranger at Arches National Monument in Utah, which led to his first book, *Desert Solitaire.* Abbey believed the American Southwest was in danger of being "over-developed" and that we must preserve the wilderness. His comic novel *The Monkey Wrench Gang* helped inspire a whole generation of environmental activism, including radical environmental warriors. His motto, taken from Walt Whitman, was "Resist much, obey little." Abbey died on 14 March 1989 at 62 and is reportedly buried in an unmarked grave in the Cabeza Prieta Desert in southern Arizona. Some claim that there is a chiseled rock with his epitaph, "No Comment."

Rachel Louise Carson (1907–1964)

Rachel Louise Carson was born on 27 May 1907 in Springdale, Pennsylvania, a small town on the western bank of the Allegheny River about 15 miles from Pittsburgh. Her father, a real estate developer, bought a large tract of land, hoping that it would become valuable when the city expanded. Unfortunately for him, the city expanded in another direction, leaving the family financially strapped. But the purchase did provide Carson with undeveloped and unspoiled countryside to explore. At the age of ten she won a prize for writing a story for *St. Nicholas*, a popular children's magazine. After high school, she enrolled in Pennsylvania College for Women (now called Chatham College) in Pittsburgh, where she continued to write. In her junior year, she changed her major from English to biology. She graduated in 1928 and did postgraduate work at the Marine Biological Laboratory in Woods Hole, Massachusetts. She received an M.A. in zoology from the University of Maryland in 1932. She went to work at the U.S. Bureau of Fisheries and in 1936 became the first female junior aquatic biologist. She continued to write, and her first book, *Under the Sea Wind,* appeared in 1941. Following the war, she helped write a series of booklets for the Fish and Wildlife Service. Her second book, *The Sea Around Us,* was published in 1951; it was on the *New York Times* bestseller list for 81 weeks.

A letter from Olga Owens Huckins changed her life. It caused her to increase her interest in the environmental effects of DDT and other chlorinated hydrocarbons. This work resulted in *Silent Spring* (1962), her third and most controversial book. The president of a chemical company that produced DDT claimed that Carson had written, "not as a scientist but rather as a fanatic defender of the balance of nature." On the other hand, U.S. Supreme Court Justice William O. Douglas called the same book "the most important chronicle of this century for the human race." If this book had not been published, it is likely that we would have to seek a new national emblem, as one of the organisms most affected by DDT was the bald eagle. Carson died on 14 April 1964 at the age of 56, of breast cancer. In 1980, Rachel Carson was posthumously awarded the highest civilian decoration in the nation, the Presidential Medal of Freedom.

George Catlin (1796–1872)

George Catlin was born in Wilkes-Barre, Pennsylvania, on 26 July 1796. He was a self-taught artist who roamed the West, painting pictures of Plains Indians and wildlife. His best-known book is the two-volume *Letters and Notes on the Manners, Customs, and Condition of the North American Indians*, in which he documented the people and landscape of the West. Catlin was one of the first to propose that the government create a "nation's park, containing man and beast, in all the wild and freshness of their nature's beauty."

J. N. (Ding) Darling (1876–1962)

Darling was a Pulitzer Prize–winning cartoonist and satirist who helped found the National Wildlife Federation. Born on 21 October 1876 in Norway, Michigan, Darling moved to Iowa as a child. It was there that he developed his interest in nature. In the early 1900s Darling embarked on a career as a cartoonist; in 1917 his cartoons were syndicated. He became active in conservation efforts, playing a role in the 1931 establishment of the Iowa Fish and Game Commission and serving as one of its first commissioners. In 1934 he was appointed chief of the Bureau of Biological Survey by President Franklin D. Roosevelt. In this capacity he was responsible for enlarging the National Wildlife Refuge System. He was president of the National Wildlife Federation (1936–1939) and lobbied for passage of the Migratory Bird Hunting Stamp Act of 1934, under which revenues from the sale of conservation stamps to hunters were used to finance wildlife refuges. Darling drew the first such stamp. A national wildlife refuge on Sanibel Island, Florida, is named for him.

John David Dingell, Jr. (1926–)

Dingell was born on 8 July 1926 in Colorado Springs, Colorado, the son of a Democratic congressman from Michigan. Dingell graduated from Georgetown University in 1949 and Georgetown University Law School in 1952. He was elected as a Democrat to the Eighty-fourth Congress to fill the vacancy caused by the death of his father in 1955. On 3 January 1973, Representative Dingell, the chairman of the House Subcommittee on Fisheries and Wildlife, introduced H.R. 37, the bill that ultimately became the Endangered Species Act. He acted as floor leader to shepherd the bill to win House approval and was an active defender of the act in 1978 during the efforts to cripple it in the wake of the controversy surrounding the snail darter. He and Representative Gerry Studds, a Democrat from Maryland, have sponsored efforts to reauthorize the act.

William O. Douglas (1898–1980)

Douglas was born in Maine, Minnesota, on 16 October 1898. He and his family moved to Yakima, Washington, where he contracted a near-fatal case of infantile paralysis in 1905. He strengthened his legs through hiking. He attended Whitman College in Walla Walla, Washington, where he graduated Phi Beta Kappa. Douglas moved to New York City to attend Columbia Law School, where one of his professors, Underhill Moore, introduced him to the ideas of Louis Brandeis. After practicing law, he returned to Columbia to teach and then moved to Yale. Because of his work on financial law, he was appointed to the Sterling Chair of Commercial and Corporate Law at Yale in 1932. He helped write the regulations for the Securities and Exchange Commission. Franklin D. Roosevelt appointed him to the Supreme Court to succeed Louis Brandeis in 1939. He was a strong supporter of wilderness preservation and conservation. In his book *A Wilderness Bill of Rights* (1965), he advocated the right to sue on behalf of nature preservation and statutory protection for wilderness areas.

David Etnier

Dr. David Etnier discovered the snail darter while snorkeling in the Tennessee River. The portion of the river was declared critical habitat for this fish, and the Tellico Dam, which was near completion, would have destroyed this habitat when it was brought on line. Construction was stopped on the dam, which led to a long series of litigation, finally reaching the Supreme Court, which found in favor of the snail darter. Ultimately, Congress passed a law that exempted the dam from the provisions of the Endangered Species Act, and President Jimmy Carter signed the law. (See Chapter 1.)

Eric Forsman

In 1968, while working for the Forest Service, Eric Forsman, a 28-year-old biology student from Oregon State University in Corvallis, had his first encounter with spotted owls. Forsman continued to observe spotted owls until he was drafted into the military in 1970. In 1972, he started working on the spotted owl in a master's program at Oregon State University. Forsman, while still a graduate student, wrote letters to the editor, gave talks at the local Audubon Society, and sent information about spotted owls to the Forest Service and the Bureau of Land Management, all in an effort to save some spotted owls he had located on the Corvallis, Oregon, city watershed. Forsman and others showed that clear-cut logging of the old growth forest habitat was causing a decline in the population of spotted owls. This work was one of the factors that led to the listing of the Northern spotted owl as a threatened species in 1990 (see Chapter 1). Dr. Forsman is a research biologist with the Forest Service.

Dian Fossey (1932–1985)

Dian Fossey was born on 16 January 1932, in San Francisco, California. She graduated from San Jose State College with a degree in occupational therapy. In 1963, she took a trip to East Africa, where she met the world-famous anthropologist Louis Leakey and had her first glimpse of mountain gorillas. As with Jane Goodall and chimpanzees, Leakey convinced Fossey of the importance of studying the great apes in their natural environment. With Leakey's help, she established the Karisoke Research Center in Parc National des Volcans in Rwanda's Virunga Mountains, one of the last spaces left to the mountain gorilla. She experienced the first known friendly gorilla-to-human contact in 1970 when Peanuts, an adult male, touched her hand. Like Goodall, she gained the trust of the gorilla bands she studied and was able to describe many previously unknown facts of gorilla behavior. When poachers killed Digit, a gorilla she had become especially attached to, she reacted by waging an increasingly vigorous public campaign against gorilla poaching, including a cover story in *National Geographic*. She established the Digit Fund to protect mountain gorillas from human activities, and she made the decision to dedicate her life to the study and protection of gorillas. She received her Ph.D. from Cambridge University in 1980 and accepted a teaching position at Cornell University. Her book *Gorillas in the Mist* was published in 1983. She returned to Karisoke to continue her campaign to ensure the survival of the mountain gorilla and to stop poaching. She was murdered at her campsite on 26 December 1985; her death remains an unsolved mystery. The last entry in her diary reads, "When you realize the value of all life, you dwell less on what is past and concentrate on the preservation of the future." The Dian Fossey Gorilla Fund was established, replacing the Digit Fund, and continues to support work at Karisoke. The mountain gorilla population is making steady gains at the Virunga Volcano area. But, there are only 650 of these magnificent creatures left in the wild. If we allow them all to disappear, the world will be a poorer place. In June 1997 the staff of the Karisoke research center suspended the work of monitoring the gorillas because of the deteriorating situation at the park.

Jane Goodall (1934–)

Jane Goodall (Baroness Jane Van Lawick-Goodall) was born in London on 3 April 1934. She attended secretarial school and held a series of jobs at Oxford University and for a documentary film studio. Goodall had always wanted to visit Africa, and when she was invited to visit a friend in Kenya, she eagerly accepted. While there, she met the world-famous paleontologist and anthropologist, Dr. Louis Leakey. Leakey encouraged her to study wild chimpanzees because their DNA differs form that of *Homo sapiens* by only about 1 percent. Leakey considered that it would take years rather than months to accumulate enough data to draw accurate conclusions about animal behavior. Goodall established a camp in the Gombe Stream Game Reserve in June 1960 to study the behavior of the chimpanzees of the region. She was one of the first scientists to make an in-depth long-term study of a primate species in its natural habitat. Her work has provided much insight into the behavior of primates, including our own species. She received a Ph.D. in ethology from the University of Cambridge, one of the few to receive a Ph.D. without having an A.B. degree. She remained at Gombe until 1975. Her books *In the Shadow of Man* (1971) and *The Chimpanzees of Gombe: Patterns of Behavior* (1986) are considered classics in the field of animal behavior.

George Bird Grinnell (1849–1938)

As a child, Grinnell moved with his family into one of the houses in Audubon Park, the estate of the late John James Audubon, author of *The Birds of America*. Grinnell studied paleontology and once accompanied General George Armstrong Custer on an expedition. Grinnell was responsible for classifying plants and animals in the West. Grinnell bought and became the publisher of *Forest and Stream* magazine. In 1886, in the pages of this magazine, Grinnell proposed an organization that would protect the nation's birds. This idea received wide acceptance, and the organization became the Audubon Society. In 1887 he and Theodore Roosevelt proposed another organization to be dedicated to ending the relentless slaughter of big game animals; it became the Boone and Crockett Club.

Ernst Heinrich Philipp August Haeckel (1834–1919)

Haeckel was born on 16 February 1834 in Potsdam, Prussia (now Germany). He studied at Wurzburg and the University of Berlin. His professor, Johannes Muller, took him on a summer expedition to the coast of the North Sea to observe small sea creatures. Haeckel completed a dissertation in zoology at Jena and was appointed to the faculty and remained there until his retirement. He was a strong supporter of Darwin and his ideas of evolution. Haecker coined the term *ecology* in 1866, from the Greek word *oikos,* for house. He used the new word to describe the study of how organisms interact with each other in a shared environment.

John Fletcher Lacey (1841–1913)

Lacey was born in New Martinsville, Virginia (now West Virginia), on 30 May 1841 and moved with his family to Iowa in 1855, where he learned the trades of bricklaying and plastering. After service in the Civil War, he was elected to the House of Representatives as a Republican for the Fifty-first Congress in 1888. He was an unsuccessful candidate for reelection, but was elected to the Fifty-third and six succeeding Congresses, serving through 1907. He was the sponsor of the Lacey Yellowstone Protection Act, which represents the first attempt by the federal government (probably the first attempt by any government, anywhere, to protect an endangered species), and the Lacey Act of 1900, which was designed to preserve, distribute, introduce, and restore game birds and other wild birds.

Aldo Leopold (1887–1948)

Aldo Leopold was born on a farm in Iowa in 1887. He attended Yale University and then joined the U.S. Forest Service. Assigned to the Arizona Territories, he began to see the land as a living organism and to develop the concept of community. In 1924, Leopold was assigned to the U.S. Forest Products Laboratory in Madison, Wisconsin, and began teaching at the University of Wisconsin in 1928. Leopold's book *Game Management* (1933) combined forestry, agriculture, biology, zoology, and ecology into a new science and defined the fundamental skills and techniques for managing and restoring wildlife populations. Soon after the publication of this book, the Department of Game Management was created at the University of Wisconsin, with Leopold as its first chair. In 1935, during the Great Depression, Leopold purchased a worn-out farm near Baraboo, Wisconsin, in an area called the sand counties. His work on this farm led to his book of essays, *A Sand County Almanac,* which is heralded as the century's literary landmark in conservation. Leopold's keen observations of the natural world are described in exceptionally poetic prose. It has guided many to discovering what it means to live in harmony with the land and with one another. This book was published posthumously in 1949. In 1948, before his work achieved public acceptance, Leopold died helping a neighbor fight a brush fire.

James Lovelock

James Lovelock redefined the concept of Gaia (from the Greek goddess of the Earth). He feels that Earth is occupied by a meta–life form that includes all of the forms of life currently existing on Earth and that they are interrelated in much the same manner as the cells and tissues of an individual organism. He states that the temperature, oxidation, state, acidity, and certain aspects of the rocks and waters are held constant by an active feedback process that operates automatically and unconsciously by the biota (the sum of all living organisms on the planet). His views were originally stated in his book, *GAIA: A New Look at Life on Earth* (1979).

Warren Grant Magnuson (1905–1989)

Magnuson was born on 12 April 1905 in Moorhead, Minnesota. He graduated from the University of Washington in 1926 and from its law school in 1929. He has served as a Democratic representative (1937–1944) and senator (1945–1980) from Washington. He was one of the key sponsors of the Endangered Species Act in the Senate. He was also one of the key sponsors of the Research or Experimentation—Cats and Dogs Act of 1966.

George Perkins Marsh (1801–1882)

Marsh was born on 15 March 1801 at Woodstock, Vermont. He was educated at Dartmouth College and was interested in silviculture and soil conservation. He was elected to Congress in 1842, where he was greatly influenced by John Quincy Adams, who had strong views on resource management and preservation. He published the first—and highly influential—textbook on conservation, including geography, ecology, and resource management: *Man and Nature, or Physical Geography as Modified by Human Action* (1864).

John Muir (1838–1914)

Muir was born in Dunbar, Scotland, on 21 April 1838. He emigrated with his family to a farm near Portage, Wisconsin, in 1849. He walked from the Middle West to the Gulf of Mexico in the 1860s. In 1868 he went to the Yosemite Valley in California and from there made many trips to Nevada, Utah, Oregon, Washington, and Alaska. He was one of the first to propose that the spectacular Yosemite formations were due to glacial erosion. Muir urged the federal government to protect unique resources such as the Sequoias and Yosemite Valley and was at least partially responsible for these areas becoming national parks. Two magazine articles by Muir helped convince the public and Congress to create 13 national forests. He accompanied President Theodore Roosevelt on a camping trip to Yosemite. Muir and Roosevelt came into conflict over building a dam in the canyon of the Tuolumne River in the Yosemite National Park to provide water to San Francisco. Roosevelt prevailed and the dam was built. Muir was one of the founders of the Sierra Club, formed in 1892 to sponsor wilderness outings in the mountain regions of the Pacific Coast. Muir was its first president (1882-1914) and soon involved the club in political activism. Muir Woods National Monument in Marin County, California, is named for the naturalist.

Gifford Pinchot (1865–1946)

Pinchot, who was born in Simsbury, Connecticut, on 11 August 1865, might be considered the "father of conservation." Pinchot graduated from Yale in 1889, and at his father's suggestion, studied forestry at the National Forestry School in Nancy, France, and in Switzerland, Germany, and Austria. After returning to the United States, he managed George Vanderbilt's Biltmore Forest. Pinchot was appointed to head the U.S. Department of Agriculture's Forestry Division by President William McKinley. But at agriculture he was a forester without a forest—the forest reserves were under the control of the U.S. Department of the Interior. In 1905, during Theodore Roosevelt's second presidential term, the control of the forest reserves were transferred to agriculture, with the creation of the Forest Service. Pinchot reportedly had a flash of insight while riding on horseback through Washington's Rock Creek Park and in his own words, "Seen in this new light, all these separate questions fitted into and made up one great central problem of the use of the earth for the good of man." Thus, Pinchot believed that using forests wisely was not an isolated issue, but that the wise use of all of the nation's natural resources formed one issue vital to the vigor and growth of the country. He initiated and served on the Public Lands Commission (1903), as well as the Inland Waterways Commission and the National Conservation Commission in 1908. He founded the Yale School of Forestry at New Haven, Connecticut, and the Yale Summer School of Forestry at Milford, Pennsylvania. He was appointed a professor of forestry at Yale in 1903 and the state forester of Pennsylvania in 1920. With his friend, Theodore Roosevelt, he advocated a utilitarian approach to conservation.

Theodore Roosevelt (1858–1919)

Roosevelt was a rancher, big-game hunter, camper, amateur entomologist, and cofounder of the Boone and Crockett Club when he ascended to the presidency in 1901 following the assassination of William McKinley. He came to the Oval Office as a true believer in conservation and a practicing environmentalist and one of the few presidents who was endowed with intellectual curiosity, as well as the ability to express himself. He moved conservation to the center of the national agenda. Conservation was a major weapon of the Progressive movement, which aimed at redressing the social, economic, and political imbalances caused by the concentration of economic power in the hands of unrestrained corporations. Roosevelt added the Yosemite Valley to the nation's national parks and launched the country's system of wildlife refuges with the acquisition of Pelican Island in Florida's Indian River. Roosevelt created the Inland Waterways Commission to study the condition of the nation's navigable waterways. In 1908 Roosevelt called the governors of all of the states to the White House for the White House Conference on Conservation, which among other things established the protection of human health as a legitimate goal of conservation. This concept is one of the central tenets of the modern environmental movement. He caused more than 196 million acres of federal lands to be closed to commercial development.

Henry David Thoreau (1817–1862)

Thoreau was born in Concord, Massachusetts, on 12 July 1817. He lived most of his life in the woods, fields, ponds, and streams surrounding his native village. Thoreau might be called the first radical environmentalist. His dictum, "In wilderness is the preservation of the world," is the battle-cry of modern environmental militants. In *A Week on the Concord and Merrimack Rivers*, he may have penned the first reference to an act of eco-sabotage, that is, an act of violence against property to protect animals and other parts of nature. He said, "I for one am with thee, and who knows what may avail a crow-bar against that Billerica Dam." The "thee" refers to the shad, whose migratory path along the Merrimack River was interrupted by the dam.

U.S. Legislation and International Treaties

4

We are caught in an inescapable network of mutuality, tied in a single garment of destiny. Whatever affects one directly, affects all indirectly.
—Martin Luther King, Jr.

U.S. Federal Legislation

Since life first appeared on Earth, species have emerged, flourished for a while, and then most have disappeared, to be replaced by a new species. When humans appeared, they began to have an impact on this process. As the number of humans increased, so did their impact. Either directly (for example, hunting, habitat destruction) or indirectly (for example, pollution), human activity has accelerated the number of species that have become extinct or are threatened with extinction. Human impact on the buffalo provides an excellent example.

It is estimated that when Europeans first began to settle North America, more than 50 million buffalo (bison) roamed the plains, prairies, open woodlands, and mountains of the North American continent. The creatures are magnificent. An adult male buffalo *(Bison bison)* stands 2.6 to 2.8 meters at the shoulder and weighs more than 1,000 kilograms. Wild canids (wolves, coyotes) can

55

take a young, old, or infirm bison, but healthy adult bison have few natural enemies in the wild—besides humans. Between the end of the American Civil War and the passage of the Yellowstone Protection Act of 1894, the number of wild buffalo was reduced from millions down to the few hundred that roamed the hills and mountains of Yellowstone National Park.

John Colter, a trapper and a member of the Lewis and Clark expedition, first described the Yellowstone area in 1808. The beauty of its mountains, canyons, lakes, and geysers was also described by mountain men such as James Bridger, Joseph Meek, and Osborne Russell, and the area was mapped by Warren Angus Ferris, a clerk with the American Fur Company, in 1836. The first official expedition to the area, in 1870, was led by General Henry D. Washburn and Lieutenant Gustavus C. Doane, who forcefully promoted the cause of establishing the area as a national park.

On 1 March 1872, Congress created Yellowstone National Park, the first extensive land area set aside by any government for the benefit and enjoyment of the people for all time. Congress mandated that "there would be no wanton destruction of the fish or game . . . capture or destruction for the purposes of merchandise or profit" and mandated "the preservation from injury or spoilage of all timber, mineral deposits, natural curiosities or wonders within said park." Unfortunately, when creating Yellowstone, Congress did not create any legal or administrative apparatus to prevent destruction of wildlife. As a result, the buffalo herd in Yellowstone, one of few remaining herds of wild buffalo, was reduced from 500 to fewer than 200 because of poaching.

Early Legislation

Lacey Yellowstone Protection Act of 1894 (28 Stat 72)

When Yellowstone National Park was created in 1872, Congress did not create a code of laws for the park, did not define offenses or punishments, and did not provide funds for administrative expenses or salaries. No law enforcement machinery was provided. As a result, poaching of wildlife, especially buffalo, was commonplace. Representative John Fletcher Lacey (R, Iowa) introduced a bill that made it illegal to remove mineral deposits, cut timber, or hunt, capture, wound, or kill game in Yellowstone. The act permitted destruction of dangerous animals when necessary to prevent them from destroying human life or causing injury. Fishing, only with hook and line, was limited to seasons and times and permitted only in a manner determined by the secretary of the interior. The secretary was also directed to promulgate rules and regulations for the park. Possession, within the park, of any dead body or parts thereof of any wild animal or bird was taken as prima facie evidence that the person was guilty of violating this act. The act allowed Yellowstone officials to confiscate any equipment or vehicles used in these offenses and provided jail terms of up to two years and fines as high as $1,000 for infractions. Finally, the act made it illegal for any railroad company, stage line, or individual to transport any game taken from the park and provided a fine of up to $300 for each offense.

The act also mandated that the United States Circuit Court appoint a commissioner, who was to reside in the park and hear and act upon all complaints of infractions of laws, rules, and regulations. The commissioner was provided with the power, upon sworn information, to issue process (for example, search warrants, arrest warrants) in the name of the United States for the arrest of any person charged with violating this act or rules and regulations promulgated by the secretary. Appeals would be handled by the United States District Court for the district of Wyoming. The commissioner was also provided with the power to issue process and cause the arrest of anyone in the park charged with the commission of a felony and provide bail according to the laws of the United States. The marshal for the district of Wyoming was directed to appoint deputies to reside in the park. The act also mandated that the secretary erect a building to be used as a jail in the park. President Grover Cleveland signed the bill into law on 7 May 1894. It was the first example of legislation to protect endangered species in the United States (and probably in the world). Unfortunately, it came about ten years after humans had hunted buffalo to virtual extinction (see David A. Dary, *The Buffalo Book* [New York: Avon Books, 1974], Chapter 8).

Lacey Act of 1900 (31 Stat 188) (see 16 U.S.C. 667[e], 16 U.S.C. 701, and 18 U.S.C. 43)

This act directed the secretary of agriculture to undertake the preservation, distribution, introduction, and restoration of game birds and other wild birds. The secretary was also directed to aid in the restoration of such birds to parts of the United States where they had become scarce or extinct, by direct purchase of birds and other means. Further, the secretary was directed to regulate the introduction of American and foreign birds or animals in localities where they had not heretofore existed. The act also mandated that the secretary collect and publish information about the propagation, the uses, and the preservation of birds covered under the law. The secretary was directed to promulgate rules and regulations to carry out the provisions of this act and to prevent the importation of wild animals or birds, except under special permits issued by the secretary. The act did not restrict the importation of specimens for natural history museums or collections or the importation of cage birds, such as domesticated canaries and parrots. The act forbid the importation of mongooses, flying foxes, fruit bats, English sparrows, starlings, and other birds and animals that the secretary determined to be potentially harmful to the interest of agriculture or horticulture. These species, on arrival at a port of the United States, were to be destroyed or returned to the owner at his or her expense. The act made it unlawful to deliver to any common carrier or for any common carrier to transport any living foreign birds or animals, any birds or animals killed in violation of the laws of the state, a territory, or the District of Columbia, or any parts of such birds or animals. All packages containing dead animals, birds, or parts thereof were to be clearly marked with the name and address of the shipper and with the contents of the package. Violators, upon conviction, were to pay a fine of $200. Sections 2 to 4 of this act were repealed by Public Law 97–79 (95 Stat 1079) and were re-enacted as sections 42 to 44 of Title 18 of the United States Code. These sections deal with game animals, including game and wild birds. Section 5 of this act was repealed.

Fish and Wildlife Conservation Act of 1966, Public Law 89–669 (80 Stat 926)

Congress found that our rapid growth and technological advance, as well as our spreading urbanization, had resulted in the extermination of some native wild animals, that the depletion of other wild species was occurring, and that the United States had obligations under certain treaties to protect and conserve our endangered species. This act was designed, in part, to provide legislative support for a comprehensive program to preserve and protect endangered species, as per our responsibilities under, for example, the Convention on Nature Protection and Wildlife Preservation in the Western Hemisphere, which the United States ratified in 1941.

The secretaries of the interior, agriculture, and defense were directed to provide a program for the conservation, protection, restoration, and propagation of selected species of native fish and wildlife, including game and nongame migratory birds, that were threatened with extinction. The act also directed the secretaries to preserve the habitats of such species "where practicable and consistent with their program purposes." A species is regarded as threatened with extinction when the secretary of the interior determines that its habitat is threatened with destruction, drastic modification, or severe curtailment because of overexploitation, disease, predation, or other factors. The secretary is directed to consult with the states and interested individuals, such as ornithologists, ichthyologists, ecologists, herpetologists, and mammalogists, and to publish a list of threatened species in the *Federal Register*. The act authorized the secretary to use Land and Water Conservation Fund Act (78 Stat 897) resources to acquire lands to protect threatened species. The secretary is directed to review the Department of the Interior's existing programs and to use those programs "where practicable" to further the purpose of the act and to consult with and assist other federal agencies to integrate the policies of the act, "where practicable." The secretary is directed to cooperate with the states and consult an affected state before acquiring any land and to enter into agreements with the states for the administration and management of any area established for the conservation, protection, restoration, and propagation of endangered species.

The act also consolidated areas administered by the secretary, such as the wildlife refuges, wildlife ranges, game ranges, wildlife management areas, and waterfowl production areas into the National Wildlife Refuge System. In administering this system, the secretary is authorized to enter into contracts with public or private agencies to provide public accommodations at locations the secretary deems are not inconsistent with the primary purposes of this act; to accept donations of funds to be used to acquire or manage lands; and to acquire lands by exchange for other acquired lands or public lands. The act makes it unlawful to disturb, injure, cut, burn, remove, destroy, or possess any real or personal property of the United States, including natural growth, in the system. The act also makes it unlawful for anyone, other than authorized personnel, to take or possess any fish, bird, mammal, or other wild vertebrate or invertebrate or any part thereof or the nest or the egg within the system. Regulations permitting hunting and fishing within the system shall be "to the extent practicable" consistent with state fish and wildlife laws and regulations. The secretary is empowered to permit use of any area in the system for hunting, fishing, public recreation, accommodation, and other activities that are compatible with the primary purpose of the act and to grant easements in, over, across, upon, through, or under any area when he or she finds that these easements are compatible with the primary purpose of the act. The act also outlines civil procedures and penalties and modifies several previous acts.

Fish or Wildlife—Endangered Species— Protection Act of 1969, Public Law 91–135 (83 Stat 275)

Congress had three purposes in passing this act. First, in Section 3, it directs the secretary of the interior to create a list of species and subspecies (geographic races or varieties) of wild mammals, birds, amphibians, reptiles, fish, mollusks, and crustaceans that are threatened with worldwide extinction. Species faced with a serious reduction in numbers in a single country are not listed. The secretary is instructed to construct this list based on the best scientific and commercial information, including that gleaned from experts in the fields of zoology, ornithology, herpetology, and ichthyology; representatives of federal and state governments; members of industries that use live animals or parts of animals; representatives of countries where the endangered species are normally found; and the International Union for the Conservation of Nature and Natural Resources.

The secretary determines whether a species or subspecies is threatened with worldwide extinction; whether its habitat is threatened with destruction, drastic modification, or severe curtailment; whether it is overutilized for commercial or sporting purposes; whether it is severely affected by disease or predation; or whether other natural or man-made factors affect its continued existence. The secretary is directed to publish the list, with scientific, common, and commercial names of species, in the *Federal Register* and regularly update it. The secretary is also directed to undertake a review of a species upon the request of interested persons. The secretary is permitted to grant limited exemptions (lasting no longer than one year) to persons importing threatened species or subspecies to prevent economic hardship and to make exceptions for zoological, educational, scientific, or propagation purposes.

The act makes it illegal to import any part, product, egg, or offspring of a listed species or subspecies into the United States.

The act does not provide for full judicial review of the secretary's determination of what species or subspecies should be included in the list. The secretary is directed to designate ports for the importation of fish (other than shellfish and fishery products) and wildlife into the United States. The secretary can grant exceptions and require an importer who imported merchandise at a non-designated port to post a bond and ship the merchandise under seal to a designated port, where the shipment can be cleared and the bond refunded. These provisions have their greatest impact on species threatened because of their commercial value and provide a stimulus for foreign governments and private groups to take more vigorous action to help protect these species. Since the act helps dry up the international market for endangered species, it should help reduce the amount of poaching of these species in their native habitat. In Section 5, the secretary is directed to encourage bilateral and multilateral agreements with foreign governments to provide protection, conservation, and propagation of fish and wildlife. Further, the secretary is directed to provide technical assistance to help develop and carry out these programs and to convene an international ministerial meeting to generate and sign a binding international convention on conservation of endangered species.

The second major purpose of this act is to help states protect domestic endangered species. The act (Section 7) amends existing law (*U.S. Code*, Title 18, Sections 43 and 44) to make the sale or purchase of any reptile, amphibian, mollusk, or crustacean or any part or egg unlawful if the sale or purchase is in violation of the laws or regulations of a state or a foreign government. The act (Section 12) extends the provisions of the Endangered Species Preservation Act (80 Stat 926) to include any reptile, amphibian, mollusk, or crustacean or any part or egg.

Third, to assist in the protection of domestic endangered species (wildlife, reptiles, amphibians, and fish), Section 12 of the act authorizes the secretary of the interior to acquire private property that is within the boundaries of any area administered (in holdings) by him or her. The purpose of this acquisition must be to conserve, protect, restore, or propagate such species. The act authorizes the secretary to spend up to $1 million per year for this purpose in fiscal years 1970 to 1972. It also increases the limit on the amount of money that can be spent on an area maintained for conservation, protection, restoration, or propagation of an endangered species of native fish or wildlife, including migratory birds, from $750,000 to $2.5 million.

The act (Section 4) outlines civil and criminal procedures, including search and seizure, arrest, forfeiture, and civil and criminal penalties. The maximum fine for a civil violation is $5,000. A criminal, or willful, violation of the provisions of the act is punishable by a fine of up to $10,000, imprisonment for not more than one year, or both. No penalties can be assessed until a person is given notice and an opportunity for a hearing. Civil actions occur in U.S. District Court. The court reviews each violation and assessment of civil penalty de novo—that is, reviews the whole process—and can reverse the decision of the secretary. The secretary is directed to keep forfeited fish and wildlife out of commercial channels, placing it with appropriate institutions for exhibition or to use for scientific or educational purposes. Section 7 of the act provides civil penalties (maximum fine $5,000) for knowingly violating the provisions of Section 4 with the exercise of due care, and the same section provides criminal penalties (maximum fine $10,000, imprisonment for not more than one year, or both) for knowingly and willfully violating the provisions of Section 4. When a person is convicted of a criminal violation, any wildlife or products thereof that were seized are automatically forfeited, as is other property seized in connection with the violation, at the discretion of the court.

The Endangered Species Act and Its Amendments

Endangered Species Act of 1973, Public Law 93–205 (87 Stat 884)

The purpose of this act is to provide a means for the conservation of threatened and endangered species and the ecosystems upon which they depend for survival, as well as to fulfill the obligations of international treaties and conventions. All federal departments and agencies are directed to utilize their authorities to conserve threatened and endangered species. Both the secretary of the interior and the secretary of commerce are directed to use all methods and procedures necessary to bring a threatened or endangered species to the point where the provisions of this act are no longer necessary. The methods and procedures of scientific resources management include, but are not limited to, research, census, law enforcement, habitat acquisition and maintenance, propagation, live trapping, and transportation, as well as regulated taking when required. The act extends protection to all members of the animal kingdom, including mammals, birds, fish, amphibians, reptiles, mollusks, crustaceans, arthropods, and other invertebrates as well as any part, product, egg, or offspring. The act excludes species of the class Insecta that are determined by the secretary to constitute a pest and whose protection would present an overwhelming risk to man. For the purposes of this act, the term *species* includes any subspecies of fish, wildlife, plants, and any other groups of fish or wildlife of the same species or smaller taxa in common spatial arrangement that interbreed when mature.

Section 4 of the act directs the secretary to determine if a species is threatened or endangered because of any of the following factors: present or threatened destruction, modification, or curtailment of its habitat or range; overutilization for commercial, sporting, scientific, or educational purposes; disease or predation; inadequacy of existing regulatory mechanisms; or other natural or man-made factors. The secretary is directed to make this determination based on the best scientific and commercial data available to him or her and after consultation with state and federal agencies and foreign countries as well as interested persons and organizations. The secretary must take several steps to list or delist a species: publish a notice in the *Federal Register*, notify the governor of each state in which the species is known to occur, allow 90 days after notification for each state to submit comments and recommendations, and publish in the *Federal Register* a summary of all comments and recommendations. The secretary is mandated to revise the list from time to time and to conduct a review of a listed species upon the request of interested persons if they present substantial evidence that a review is warranted. Furthermore, the secretary is directed to issue regulations that provide for the conservation of a listed species. The secretary must publish the proposed regulation in the *Federal Register* at least 60 days before the effective date of the regulation. Interested individuals may raise an objection and request a public hearing. If the secretary denies a public hearing, he or she must publish the reasons in the *Federal Register*. The secretary may also treat a species as threatened or endangered, even though it is not listed, if it closely resembles a species that is listed and if treating the unlisted species as endangered will substantially facilitate the enforcement of regulations concerning the listed species. The secretary must take into account efforts being made by any nation or any political subdivision of a nation to protect species.

The act (Section 5) directs the secretary to acquire real property (land, waters, or interests therein) by purchase, donation, or otherwise if he or she finds it necessary to do so to conserve, protect, restore, or propagate threatened or endangered species. He or she uses the authority granted under the Fish and Wildlife Act of 1956, the Fish and Wildlife Coordination Act, and the Migratory Bird Conservation Act to acquire this real property and uses funds made available by the Land and Water Conservation Fund Act of 1965.

Section 6 of the act directs the secretary to cooperate with the states to the maximum extent possible. He or she is required to consult with a state before acquiring any real property in that state, to enter into management agreements for any area established for protective purposes, and to provide financial assistance, in the form of grants, for threatened or endangered species programs. The act voids any state law that is less restrictive than the act but does not void any other state law or regulation intended to conserve migratory, resident, or introduced fish or wildlife or to permit or prohibit sale of such fish or wildlife. State law may be more restrictive than the act.

In Section 7, the secretary is directed to review all programs administered by him or her and to utilize these programs to further the purpose of this act. All other federal departments and agencies are also directed to utilize their authorities for the furtherance of the purposes of this act, and these departments and agencies are mandated to ensure that any actions authorized, funded, or carried out by them does not jeopardize the continued existence of threatened or endangered species or result in destruction or modification of habitat of a threatened or endangered species.

The president (Section 8) may use foreign currencies accruing to the United States government to provide any foreign country (with its consent) with assistance in developing and managing conservation programs that are necessary and useful for the conservation of any listed species. The president is also authorized and directed to designate agencies to act as management authority and scientific authority under the Convention on International Trade in Endangered Species of Wild Fauna and Flora (see below); their duties include the issuance of permits and certificates. The president also designates agencies to act on behalf of the United States government with respect to the Convention on Nature Protection and Wildlife Preservation in the Western Hemisphere (see below).

Prohibited acts (Section 9) for any person subject to the jurisdiction of the United States include importing into or exporting from the United States any listed species; taking (harassing, harming, pursuing, hunting, shooting, wounding, killing, trapping, capturing, or collecting or attempting to engage in such conduct) any listed species in the United States; taking any listed species on the high seas; possessing, selling, delivering, carrying, transporting, or shipping by any means whatsoever any species taken in violation of the act; delivering, receiving, carrying, transporting, or shipping in interstate or foreign commerce by any means whatsoever in the course of commercial activity any listed species; selling or offering for sale in interstate or foreign commerce any listed species; or violating any regulation pertaining to such species. It is also unlawful for any person to engage in any trade in any specimens contrary to the provisions of the Convention on International Trade in Endangered Species of Wild Fauna and Flora. Any importer or exporter of fish, wildlife, or plants must obtain the permission of the secretary, maintain records and file reports as mandated by the secretary, and allow representatives of the secretary to examine his or her inventory of imported fish, wildlife, or plants. It is also unlawful to import or export fish, wildlife, or plants except at designated ports.

The secretary (Section 10) may issue permits to engage in acts otherwise prohibited by Section 9 for scientific purposes or to further the propagation or survival of a listed species. The secretary may also provide limited exceptions to prevent economic hardship. Alaskan natives (Indian, Aleut, Eskimo, or permanent nonnative residents of an Alaskan native village) may take threatened or endangered species for subsistence purposes, provided that the taking is done in a nonwasteful manner.

The act (Section 11) provides civil and criminal procedures and penalties for violations. Under Section 11g, any person may commence a civil suit to enjoin any person, including the United States and any governmental instrumentality or agency, who is alleged to be in violation of any provision of this act or regulations issued under the authority of this act. Anyone may also compel the secretary to prevent the taking of any resident threatened or endangered species.

Endangered Species Act—1976 Amendments, Public Law 94–359 (90 Stat 911)

The Endangered Species Act of 1973 prohibits the importation and sale of endangered species and their parts and products in interstate and foreign commerce. The 1973 act did not make any retroactive exceptions for the interstate sale of parts and products of endangered marine species that were legally held under the provisions of the Marine Mammal Protection Act of 1972. The 1976 amendments to the Endangered Species Act of 1973 deal with scrimshaw and whale oil.

New England whalers originated scrimshaw in the nation's early years. The whalers etched designs, usually nautical motifs, on whale (Cetacea) bone or teeth. The whalers also carved figures of the material. As the nation spread westward, this art form also spread to areas such as the Pacific Northwest, the Southwest, Alaska, and Hawaii. Today's artisans have adapted the art form to other nonnautical motifs. The 1976 amendments to the Endangered Species Act exempts from the provisions of the act any finished scrimshaw product or raw material for such a product that was lawfully held prior to December 20, 1973, and allows exportation and interstate sale of these items. This amendment minimizes the economic hardship caused by the act on scrimshaw artisans and helps preserve scrimshaw as an art form. A person seeking an exemption must apply for the exception within one year of when the amendments were enacted, and the application must include a detailed inventory of endangered species parts held before the act. The secretary of commerce cannot grant a permit for more than three years, so all scrimshaw must be sold or otherwise disposed of within three years.

From 1948 to 1952, under the authority of the Strategic and Critical Materials Stockpiling Act, the General Services Administration acquired and stockpiled sperm whale oil, which was used as a lubricant. In 1972, the Interdepartmental Materials Advisory Committee determined that effective substitutes for sperm whale oil existed and that sperm whale oil should no longer be considered a strategic material. The 1976 amendments to the Endangered Species Act allow the General Services Administration to dispose of the surplus sperm whale oil.

Endangered Species Act—Amendments of 1978, Public Law 95–262 (92 Stat 3751)

The 1978 amendments to the Endangered Species Act require all federal agencies to consult with the Fish and Wildlife Service or the National Marine Fisheries Service to determine if a proposed action might have an adverse impact on an endangered or threatened species. The amendments also mandated the establishment of an Endangered Species Committee that would consist of a minimum of seven members: the secretary of the interior (who is the chair), the secretaries of agriculture and the army, the chair of the Council of Economic Advisors, and the administrators of the Environmental Protection Agency and the National Oceanic and Atmospheric Administration, as well as one individual from each affected state, appointed by the president. The committee was to provide flexibility in dealing with controversies arising between a federal project and an endangered species, as covered by Section 7 of the Endangered Species Act of 1973. Five members of the committee constitute a quorum. The committee meets when called by the chair or five of its members, and all records of meetings are open to the public. The committee can request the help of personnel of any federal agency to achieve its goals; can obtain from any federal agency, subject to the Privacy Act, any information it requires to carry out its duties; and can hold hearings to take testimony and evidence as it deems advisable. The committee can promulgate and amend rules, regulations, and procedures and can issue and amend orders as it deems necessary, as well as subpoena witnesses for attendance and testimony and subpoena relevant papers, books, and documents. The committee members may designate a person to act as a representative if the person holds a federal office that is filled on the advice and consent of the Senate. The representative cannot cast a vote on behalf of any member.

The secretary of the interior is mandated to promulgate regulations that set forth the form and manner in which applications may be made for exemptions from the provisions of the Endangered Species Act of 1973. These applications can be submitted by any federal agency, the governor of any state, or any applicant for a permit or a license. In the case of a federal agency, the application must detail what consultation process was carried out between the head of the agency and the secretary of the interior and must include a statement describing why such action cannot be altered or modified to conform with the requirements of the act and its amendments. The application, which must be filed within 90 days of the completion of the consultation process, is considered by a review board, which reports its findings to the committee. The committee makes the final determination.

The board consists of an individual appointed by the secretary, an individual who resides in the state in which agency activity will be or is being carried out, appointed by the president, and an administrative law judge. The secretary is directed to submit the application to the board and provide the board with his or her views and recommendations in writing. The board must decide by majority vote whether an irresolvable conflict exists and whether the applicant has carried out its consultation responsibilities, conducted any biological assessment, and refrained from making any irreversible or irretrievable commitment of resources. If the board decides that an irresolvable conflict does not exist or that the applicant has not met the conditions mentioned above, the agency action is considered final. If the board finds that an irresolvable conflict does exist and the applicant has met the conditions, then the board must, within 180 days, report to the committee on the availability of reasonable and prudent alternatives to the agency action consistent with conserving the species or its critical habitat.

The board must also present the committee with a summary statement of evidence of the availability of reasonable and prudent alternatives to the agency action and on the nature and benefits of agency actions and alternatives that are consistent with conserving species and critical habitats. In House Report 95 1625, the Subcommittee on Fisheries and Wildlife Conservation and the Environment of the Committee on Merchant Marine and Fisheries held that both ecological and economic considerations should be kept in mind when determining whether an exemption should be granted. The subcommittee indicated that the board should consider factors not specifically called out in the act, including the cost and its impact on consumers, business markets, and federal, state, and local governments; productivity; competition; employment; energy supply and demand; and supplies of important materials, products, and services. Further, the board must provide a summary of the evidence as to whether the agency action is in the public interest or is of national or regional significance and as to what appropriate mitigation and enhancement should be considered by the committee. The actions must be reasonable in their cost, their likelihood of protecting the species, and the availability of technology required to make them effective. They must seek to minimize the adverse effects of agency action on the endangered or threatened species. They can include, but are not limited to, live propagation, transplantation, and habitat acquisition and improvement. The board must collect this evidence at a formal adjudicatory hearing. The board can take testimony and receive evidence; request information, subject to the Privacy Act, from any federal agency; request personnel from any federal agency; and use the mails in the same manner and under the same conditions as a federal agency can. The committee must make a final determination about whether to grant an exemption or not within 90 days of receiving the report of the board. It grants an exception based on the vote of five of its members voting in person if it determines based on the report of the board and any evidence it acquired that there is no reasonable and prudent alternative to the agency action. The benefits of the action must clearly outweigh the benefits of alternative courses of action that are consistent with conserving the species or its critical habitat. The action must also be of regional or national significance, and the action must be in the best interests of the public. The committee is prohibited from considering an exemption if the secretary of state certifies that granting an exemption and carrying out the act would be in violation of an international treaty or other obligation of the United States. Two other exceptions are authorized. First, the Endangered Species Committee must grant any exemption called for by the secretary of defense, provided that the exemption is necessary for reasons of national security. Second, the president is authorized to grant exemptions for the repair and replacement of a public facility in any area declared to be affected by a major disaster.

If the committee, after its review, decides that an exemption should be granted, the committee issues an order granting the exception and specifies the mitigation and enhancement measures that the exemption applicant must carry out and pay for. The applicant must provide yearly reports to the Council on Environmental Quality describing its compliance with these requirements until they are completed. Any person can obtain a judicial review of any decision of the committee in the United States Court of Appeals.

The secretary is also directed to publish any proposed regulation (including the entire text of the regulation) not less than 60 days before its effective date in the *Federal Register,* in a scientific journal, and in a newspaper of general circulation. The regulation and any environmental statements must be presented to all general local governments within or adjacent to a critical habitat, if one is designated. The secretary is also directed to hold a public meeting within or adjacent to the critical habitat. He or she must consider the economic and other impacts of specifying a critical habitat. The secretary can exclude an area from the critical habitat if he or she determines that the benefits of the exclusion outweigh the benefits of including the area, unless he or she finds that the noninclusion will result in the extinction of the species. The secretary must also publish in the *Federal Register* a brief description and evaluation of public or private activities that might adversely modify the habitat or be impacted by the designation of a critical habitat. The act states that a final regulation adding a species to the endangered or threatened species list must occur within two years after the initial publication of the proposed regulation. If that deadline is not met, the secretary must withdraw the regulation.

The secretary is directed to develop and implement a recovery plan for the conservation and survival of an endangered or threatened species. The secretary may procure the services of public and private agencies, institutions, and qualified individuals to help implement such a plan and to appoint a recovery team to implement the plan.

The amendments authorize appropriations to be used in the furtherance of the purposes of the act and to assist the committee and review boards in carrying out their functions. The amendments also make modifications in civil and criminal procedures and penalties. The act also makes changes to existing laws with respect to raptors (predatory birds) and antique articles made from parts of endangered species. It also directs the committee to consider the Tellico Dam and Grayrocks Dam and Reservoir projects, as well as the Missouri Basin Power Project to determine if they should be exempt from Section 7(a) of the Endangered Species Act. The Tellico Dam later became the subject of considerable litigation (see Chapter 5) and would ultimately be exempted by the committee. The act also authorizes the secretary to enter into cooperative agreements with the states and to assist states in programs for the conservation of endangered species of animals and plants. The act mandates that at the time of listing an endangered or threatened species, the secretary, to the maximum extent prudent, specify the critical habitat of the species. The secretary is also directed to review the endangered and threatened species lists and to determine the status of each species.

Authorization, Appropriations—Endangered Species Act of 1973, Public Law 96–159 (93 Stat 1225), as passed in 1979

The main function of this act is to provide funding for the Endangered Species Committee, the review boards, and the Department of the Interior to carry out the provisions of the Endangered Species Act of 1973 (described earlier in this chapter) through the end of fiscal year 1982. This act extends protection to plants and mandates that the secretary conduct a review of the status of a species before proposing to list it. The review should include consultations with experts in the field (such as regional, area, and field staff, university professors, representatives of professional organizations, local citizens, and representative of state and federal agencies) and a perusal of professional journals. The act's requirement for local publication is modified to require a summary of the listing proposal, rather than the full text. The summary should be published in a newspaper of general circulation within or adjacent to the habitat and should include the biological justification for the listing and the justification for the critical habitat designation, as well as a brief description of the activities that might be adversely affected by the designation of the habitat and what activities might adversely modify the critical habitat. A map of the critical habitat must also be published. Public meetings (informal exchange of information on regulatory proposal) and public hearings (more formal opportunities for residents to comment on the listing and designation proposal, with verbatim transcripts for the record) are to be held separately.

The secretary is directed to establish and publish in the *Federal Register* agency guidelines that include procedures for recording the receipt and the disposition of petitions submitted for exemption, making findings, ranking species as to which should receive priority review, and developing and implementing, on a priority basis, recovery plans.

This act makes technical changes in the existing law to clarify congressional intent on the exception process to nonfederal applicants and also clarifies the wording of Section 7 of the Endangered Species Act of 1973 to bring the language of the statute into conformity with agency practice and recent federal court decisions, such as *National Wildlife Federation v. Coleman* and *Tennessee Valley Authority v. Hill* (see Chapter 5). This act mandates that the secretary render biological opinions based on the best evidence available. If the evidence is inadequate, the act directs that the federal agency has a continuing obligation to make a reasonable effort to develop additional information. Furthermore, if federal agencies proceed with a potentially threatening project on the basis of inadequate information, they run the risk that the project might be stopped in the future as more adequate information about the species becomes available. Federal agencies are also directed to have informal discussions with the secretary about actions that might jeopardize any species that is proposed to be listed as threatened or endangered. Individuals and organizations that seek a permanent exemption may conduct a biological assessment (under the supervision of the secretary and the appropriate federal agency) to identify threatened or endangered species that might be affected by their actions.

The act establishes the secretary of the interior as the scientific and management authority for the purpose of the Convention on International Trade in Endangered Species of Wild Fauna and Flora (see below). The secretary is directed to establish a seven-member International Convention Advisory Commission. It includes members appointed by the secretaries of the interior, agriculture, and commerce, the director of the National Science Foundation, the chair of the Council on Environmental Quality, and a representative of the Smithsonian Institution, as appointed by its director. The seventh member is chosen by the secretary from officers and employees of the state agencies that have fish and wildlife conservation and management responsibilities. The members must be scientifically qualified and are appointed for two-year terms. The commissioners elect a chair from among their members, and all decisions of the commission must be made by majority vote. The commission reports its recommendations to the secretary and must include any written dissenting view made by a member. In discharging its responsibilities, the commission must consult with experts in the scientific community. If the secretary disagrees with any majority decision of the commission, he or she is required to provide a written explanation to the commission and to publish the findings of the commission and his or her decision in the *Federal Register*. The chair appoints, with the concurrence of the commission, an executive secretary to carry out the duties and functions prescribed by the commission. The president designates the agencies of the federal government that will act on behalf of and represent the United States as required by the Convention on Nature Protection and Wildlife Preservation in the Western Hemisphere. The act allows owners of legal sperm whale oil and scrimshaw a three-year extension of the exemption created by the 1976 amendments to the Endangered Species Act (described previously in this chapter).

Endangered Species Act Amendments of 1982, Public Law 97–304 (96 Stat 1411)

These amendments authorized appropriations for fiscal years 1983, 1984, and 1985 to carry on Endangered Species Act activities and provided additional appropriations to carry out the requirements of the Convention on Nature Protection and Wildlife Preservation in the Western Hemisphere.

The amendments mandate that the secretary must, to the maximum extent prudent and determinable, designate critical habitat at the time a species is listed. This means that if the biological status of a species is clear, it should not be denied the protection of the act because of the additional (economic) studies required to establish a critical habitat. The secretary is to make his or her determinations for listing based "solely" on the basis of the best scientific and commercial data available and not on economic considerations. Designation of critical habitat must be made on the basis of the best scientific data available but must take into account the economic and other impact of specifying a particular area part of the critical habitat.

To the maximum extent possible, the secretary is required to make a determination, within 90 days, of whether a petition for listing or delisting a species (or modifying a critical habitat) presents substantial scientific or commercial information indicating that the petitioned action is warranted. If action is warranted, the secretary is required to begin a review of the status of the species and publish his or her findings in the *Federal Register*. Within 12 months after receiving a petition, the secretary must determine the final status of the species—that is, whether the species should be listed or delisted on the basis of the information presented in the petition and the findings of the secretary—and publish his or her findings in the *Federal Register*. If action is warranted, the secretary must publish a notice in the *Federal Register* and complete the text of the proposed regulation. The designation of critical habitat should not delay this process, but critical habitat must be specified within one year of listing. In an emergency, the secretary has the discretion to make the regulations effective immediately upon publication in the *Federal Register*. If the normal rule-making procedures are followed within 240 days, they remain in effect. If the normal procedures are not followed the regulation expires at the end of the 240 days.

The amendments increase the federal share of grants to states from 66.66 percent to 75 percent for single-state projects and from 75 percent to 90 percent for multistate projects.

A federal agency is required to consult with the secretary on any prospective agency action at the request of and in cooperation with a permit or license applicant for any activity for which the secretary has issued a biological opinion. The agency must also inform the applicant whether it is likely that the permit will be granted or not. The secretary is to give priority to the preparation of recovery plans that are or may be in conflict with construction or other development projects. Early consultation is encouraged and hopefully involves the secretary as well as state and local planning agencies and conservation entities. If so warranted, the secretary must provide a written statement detailing how the proposed action will jeopardize endangered or threatened species or result in the destruction or adverse modification of critical habitat. If this is the case, the secretary is required to suggest reasonable and prudent alternatives that will minimize these effects. Incidental taking of species is allowed under the terms and conditions specified by the secretary to minimize the impact of these takings. If the specified impact on the species is exceeded, the agency must seek an additional consultation with the secretary but is not required to cease all operations unless it is clear that additional takings will cause irreversible or adverse impact on the species.

Agencies can seek an exemption from the act. The three-member review board that determines whether an irresolvable conflict exists and whether the agency carried out its consultation responsibilities in good faith is eliminated, and the functions of the review board are carried out by the secretary. The secretary must make the initial determination of whether irresolvable conflicts exist within 30 days, and the report discussing reasonable and prudent alternatives must be completed within 120 days. The secretary must consult with members of the Endangered Species Committee. After receiving the report of the secretary, the Endangered Species Committee has 30 days to reach a decision.

The secretary is required to make export determinations and give advice on exports based on the best available biological information and professionally accepted practices in wildlife management. The secretary is not required to make nor is he or she allowed to have any state make estimates of population size in calculating export determinations or in giving such advice. Provided that no accepted wildlife management practices determine that export will be detrimental to the survival of the species, then exports will not be disallowed merely because population estimates are not available. The amendments abolish the International Convention Advisory Committee. They also provide that if the United States votes against including any species in Appendix I or II of the Convention on International Trade in Endangered Species of Wild Fauna and Flora (CITES) or does not enter a reservation pursuant to Article XV of the convention, the secretary of state will provide Congress with a written report specifying the reason why a reservation was not entered.

The amendments implement activities associated with the Convention on Nature Protection and Wildlife Preservation in the Western Hemisphere. The secretary is directed to cooperate with contracting parties to develop resources and programs, identify species of birds that migrate between the United States and other contracting parties, and implement cooperative measures to ensure that these species do not become endangered or threatened. The secretary is also to implement provisions that address protection of wild plants.

The amendments establish a procedure that enables the secretary to issue permits to allow incidental taking—that is, the taking of members of listed species that is incidental to an activity and where the taking is not the purpose of the activity. The applicant for a permit must provide the secretary with a plan that specifies the number of the species likely to be taken, the steps the applicant will take to minimize the takings, and what alternative actions were analyzed and why they were not adopted.

The amendments extend the exemption from trade restrictions for finished scrimshaw products that was authorized in the 1976 amendments and renewed in the 1979 amendments. This extension applies only to finished scrimshaw and not to raw whalebone or teeth. The exemption is extended to antique scrimshaw made 100 years before the date of importation.

The amendments define an experimental population as a population of an endangered or threatened species authorized by the secretary for release of eggs, propagules, or individuals that the secretary determines will further the conservation of the species. The experimental population is wholly separate geographically from the nonexperimental populations of the same species. A critical habitat is not designated for experimental populations. Any experimental population found in any unit of the National Wildlife Refuge System or the National Park System is subject to full protection under Section 7 of the Endangered Species Act.

Title I Endangered Species Act Amendments of 1988

The 1988 amendments authorize appropriations through the end of fiscal year 1992. The amendments grant the U.S. Fish and Wildlife Service enforcement authority over the importation and exportation of plants protected by the Endangered Species Act and under the Convention on International Trade in Endangered Species of Wild Fauna and Flora (CITES). Until this point, that authority was vested solely in the Animal and Plant Health Inspection Service of the Department of Agriculture. The act mandates that the secretary implement a system to monitor the status of all candidate species and grants the secretary the authority to implement emergency listings if this will prevent significant risk to the well-being of the species. The U.S. Fish and Wildlife Service must publish and periodically update comprehensive notices containing lists of native species being considered for listing. The service must consult, coordinate, and encourage communication with other federal and state agencies, private conservation organizations, and members of the academic and scientific communities in developing these lists. The service need not regulate both trade and taking of species listed as threatened or endangered because of their similarity of appearance to other listed species if regulation of only one of these activities is sufficient to protect the species.

The act mandates the development of recovery plans without regard to the species' taxonomic classification, especially for species that are or may conflict with development projects or other forms of economic activity. Each plan must include a description of site-specific management actions that are necessary to achieve the plan's goal, as well as objective, measurable criteria, which, when met, allow delisting of the species. The plan should contain estimates of the amount of time required and the cost to carry out the plan. The secretary can acquire the services of appropriate public and private agencies, institutions, and individuals. The secretary is also directed to implement a system to monitor species that have been recovered and that have been delisted for five years.

The secretary is authorized to provide financial assistance to any state so that it can enter into cooperative agreements to develop programs for the conservation of endangered and threatened species.

The act makes it unlawful to remove and reduce to possession any species of plant from areas under federal jurisdiction, to maliciously damage or destroy any such species in any such area, or to cut, dig up, or damage any such species.

The secretary of commerce is directed to contract for an independent review of scientific information pertaining to the conservation of each relevant species of sea turtles, especially with regard to minimizing mortality by shrimp trawling. The study is to be conducted by the National Academy of Science using experts not employed by federal or state government, except employees of state universities, who have knowledge of sea turtles and activities that might adversely affect them. The study is also to identify appropriate conservation and recovery measures as well as reproductive measures. The report is to be presented to the Committee on Environment and Public Works for the U.S. Senate, the Committee on Merchant Marine and Fisheries of the House of Representatives, and the secretary on or before 1 April 1989. On receipt of the review, the secretary must review the status of each relevant species of sea turtle and determine if the regulations promulgated on 29 June 1987 need to be modified or enforced as written. The secretary is directed to undertake educational efforts among shrimp trawlers in the usage of the turtle excluder and any other device that might be imposed.

The administrator of the Environmental Protection Agency and the secretaries of agriculture and the interior must conduct a program to inform and educate persons engaged in agricultural food and fiber production about proposed labeling programs or requirements that may be imposed by the administrator in compliance with the Endangered Species Act. The program is to identify pesticides affected by the program, the geographic areas affected by any pesticide restriction or prohibition, and the effects of the restricted pesticides on endangered or threatened species.

Title II—African Elephant Conservation Act, Public Law 100–478 (102 Stat 2306)

The elephant populations of Africa have declined at a high rate since the 1970s, mainly due to illegal trade in African elephant ivory. The African elephant was listed as threatened under the Endangered Species Act, and the Asian elephant is listed as endangered. The secretary of the interior may provide financial assistance from the African Elephant Conservation Fund for approved projects, including research, conservation, management, and protection of African elephants. Any individual with experience and expertise in African elephant conservation may submit a project. The secretary must review each project to determine if it meets the appropriate criteria.

The African Elephant Conservation Fund is established, under Title II, as a general fund of the treasury. The secretary of the treasury must deposit into the fund all amounts received in the form of penalties, donations, and other amounts appropriated to the fund by Congress. The secretary must submit an annual report to Congress on the status of the African elephant.

The secretary calls for information about the African elephant conservation program from each ivory-producing country. The secretary evaluates the conservation program of each ivory-producing country and determines whether the country is party to the Convention on International Trade in Endangered Species of Wild Fauna and Flora (CITES) and adheres to the CITES Ivory Control System, whether the country's elephant conservation program is based on the best available information, whether the country is making progress on compiling information about its elephant population, whether the taking of elephants is controlled and monitored, and whether the country's ivory quota is determined on the information obtained above. If a country does not meet these criteria, the secretary establishes a moratorium on importation of raw and worked ivory. The secretary must establish a moratorium on the importation of raw and worked ivory from an intermediary country if it is not a party to CITES, if it imports raw ivory from a country that is not an ivory-producing country, or if it imports raw or worked ivory from a country that is not a party to CITES or from a country that is under a moratorium. The moratorium is suspended when the secretary determines that the reasons for establishing the moratorium no longer exist. Individuals may import sport-hunted elephant trophies that are legally taken in an ivory-producing country.

The act makes it illegal for any person to import raw ivory from any country other than an ivory-producing country, to export raw ivory from the United States, to import raw or worked ivory that was exported from an ivory-producing country in violation of its laws or the CITES Ivory Control Program, or to import worked ivory, other than personal effects, from a noncertified country or one that is under a moratorium. The act outlines criminal and civil penalties, as well as rewards.

Recent Legislative History

The authorization of the Endangered Species Act expired on 30 September 1992. The Endangered Species Act Amendments of 1988, voted on by the Senate on 15 September 1988 and the House of Representatives on 26 September 1988 and signed into law by President Ronald Reagan on 7 October 1988, were the source of appropriated funds for the act's implementation through fiscal year 1992. Implementation of the act continued via annual appropriations for the Departments of the Interior and Commerce. In February 1995 Congressman Don Young (R,

Alaska), the chair of the House Resources Committee, announced the formation of the House Endangered Species Act Task Force, chaired by Congressman Richard Pombo (R, California). This task force held public meetings during 1995.

On 23 February 1995, during consideration of H.R. 450, the Regulatory Transition Act, the House of Representatives agreed by voice vote to stop all listings under the Endangered Species Act until it is reauthorized. On 16 March, during consideration of the fiscal year 1995 Defense Supplemental Appropriation Bill (H.R. 889), the Senate supported an amendment to prohibit the listing of addition species under the Endangered Species Act until the end of 1995 or until the act is reauthorized. President Bill Clinton signed the bill into law on 10 April 1995.

The House of Representatives and the Senate, on 18 July and 9 August 1995, respectively, extended the moratorium on listing and prelisting species until September 1996 or until the act is reauthorized.

Congressmen Don Young and Richard Pombo and a bipartisan group of 116 cosponsors introduced H.R. 2275, the Endangered Species Conservation and Management Act. Senator Dirk Kempthorne (R, Idaho), the chairman of the Senate Subcommittee on Drinking Water, Fisheries, and Wildlife of the Environment and Public Works Committee, introduced S 1364, the Endangered Species Conservation Act.

On 26 April 1996 President Clinton lifted the moratorium on listing and prelisting. Congress granted the president this power based on his threat to veto the FY 1996 funding bill for the Department of the Interior and other segments of the government.

Representative Young introduced the New Wildlife Refuge Authorization Act on 4 February 1997. It was reported out of the Committee on Resources on 29 September 1997 and the Rules Committee on 12 May 1998.

Representative Saxton introduced the Asian Elephant Conservation Act of 1997 (S 1287) on 4 June 1997. This bill provided assistance for the conservation of Asian elephants by supporting and providing financial resources for the conservation programs of nations within the range of Asian elephants and providing support to those persons with expertise in the conservation of Asian elephants. This bill was passed by voice vote by the House on 21 October 1997 and in the Senate by unanimous consent on 8 November 1997. It was signed into the law by the president on 19 November 1997, becoming Public Law 105-96.

Senators Dirk Kempthorne (R, Idaho) and John Chafee (R,

Rhode Island) released a discussion draft of their proposed ESA reform legislation on 1 January 1997.

Representative Bills

These bills were all in early stages of the legislative process in 1998. It seems unlikely that any of these bills will be enacted in their present form, or if enacted, will be signed by the president. The first, the Endangered Natural Heritage Act, is a draft bill developed by a coalition of individuals and organizations interested in the Endangered Species Act. It has yet to be introduced into Congress.

Endangered Natural Heritage Act (ENHA)

The framers of the ENHA believe that a species is more likely to recover if its population is big enough to withstand genetic, environmental, and natural obstacles. Therefore, the ENHA would cause federal agencies to act proactively to conserve declining species before they need the full protections of the Endangered Species Act. The ENHA would mandate that the Fish and Wildlife Service (FWS) and the National Marine Fisheries Service (NMFS) make a decision whether or not to list a candidate species within four years of its being proposed for listing. The act would also establish a National Commission on Species Extinction to survey and identify species at risk—including indicator species, umbrella species, and species about which little is known—as well as imperiled ecosystems to allow federal and state agencies to set priorities and develop multispecies and ecosystem strategies. Because many rare species occur on private lands, the commission would set up cooperative programs with private landowners to arrange for private land surveys.

Under the ENHA, plants would be afforded the same protection as animals. The ENHA would direct that essential habitat be identified at the time that a species is listed and that characterization of the critical habitat be refined and finalized in the final recovery plan for the species. The ENHA would limit the current "not prudent" or "not determinable" exceptions to critical habitat designation and make it clear that current Endangered Species Act requirements to take economic impact into consideration should not be used to delay the designation of critical habitat. The ENHA makes it explicit that "destruction or adverse modification" of critical habitat means any action that diminishes the value of critical habitat for recovery of a listed species in the wild.

The Endangered Species Act does not currently have an explicit requirement that federal agencies implement recovery plans nor are plans typically detailed enough to clearly establish if they are being followed. The ENHA would require that objective scientific benchmarks be established so that listing, recovery, and delisting decisions utilize all available information on population size and distribution, specific critical habitat requirements, and other relevant biological criteria. The ENHA also stipulates that state-of-the-art quantitative modeling techniques be used whenever possible. The ENHA would require the secretary of the interior to prioritize actions so they (1) have the greatest potential for achieving recovery of listed species, (2) avoid imminent extinction, and (3) benefit the largest number of species. The act would also require the secretary to identify in the recovery plan types of actions likely to hurt the species and violate the act. The ENHA would require the secretary to release draft recovery plans within 18 months of listing and to finalize the recovery plans within 12 more months, after public review and comment and after deadlines for implementation of recovery objectives are established. Significantly affected federal agencies would be required to finalize agency implementation plans within 6 months of recovery plan finalization.

The ENHA would limit or eliminate "no effect" determinations through "informal" Endangered Species Act consultations by requiring documentation and public access to records of all formal and informal consultations and by soliciting public comment on draft biological opinions. Habitat Conservation Plans and incidental taking would be permitted only after a scientific and quantitative evaluation of the impact on the species determines that they will not negatively impact recovery. The ENHA would enhance citizen enforcement of the Endangered Species Act and increase civil and criminal penalties for violations of the act.

Endangered Species Conservation and Management Act of 1995 (H.R. 2275)

If any action under the Endangered Species Act diminishes the value of any portion of real or personal property or water rights by 20 percent or more, H.R. 2275, introduced by Representatives Don Young (R, Alaska) and Richard Pombo (R, California) on 7 September 1995, would require the federal government to pay compensation to the property owner. The federal government, at the property owner's insistence, must buy the portion of the property affected, paying fair market value based on the value of the property before the diminution occurred, if the diminution in value is greater than 50 percent. Owners, filing a written request within one year of receiving notice of the property limitations, may elect binding arbitration or may file a civil suit to obtain compensation. Compensation is paid out of the annual appropriation of the agency taking the action and includes payment of the owner's attorney fees and costs. The secretary of the interior is authorized to provide grants to private property owners if the property contains habitat that significantly contributes to the protection of the population of a listed species. The secretary is required to establish technical assistance programs to provide public information on habitat management, species propagation, feeding, predator control, and any other topics that a nonfederal person—that is, an individual, organization, or corporation that is not part of the federal government—may require to conserve a listed or candidate species.

H.R. 2275 would allow any nonfederal activity that is consistent with a conservation plan or objective, that complies with an incidental take permit or a cooperative management agreement, or that addresses a critical, imminent threat to public health or safety. The act would limit federal authority to seize fish, wildlife, or plants or to require forfeiture of gear and equipment used in illegally taking threatened and endangered species. The act would also mandate that all measures to implement and enforce the Endangered Species Act be promulgated through rule making, with public notice and comment; multiple analyses of the impacts of the rule; a summary of the literature reviewed, the experts consulted, and the secretary's findings based on that review and consultation; and an analysis showing that compliance with the rule is reasonably within the means of the entity involved.

H.R. 2275 authorizes the secretary to enter into cooperative agreements with state and local governments and nonfederal persons for the management of any listed or candidate species or its habitat. Private and other nonfederal property cannot be restricted under a cooperative management agreement unless the owner of the property consents in writing. The secretary is authorized to provide funds to a nonfederal person to carry out the agreement. The secretary must provide public notice and public hearing on proposed cooperative management agreements and to approve an agreement within 120 days. The secretary is to notify any party that is not fulfilling the agreement, and if the party fails to take corrective action within 90 days; the secretary must rescind the agreement.

H.R. 2275 would limit citizen suits to actions to enjoin federal agencies and officials from violations that pose a threat of immediate and irreparable harm to a listed species. Citizens could also sue to compel the secretary to enforce or modify the prohibition against taking listed species, to apply or modify permit provisions, or to perform any nondiscretionary duty to issue regulations governing the conservation of threatened species. H.R. 2275 would also allow citizen suits by persons claiming economic or other injury.

The act eliminates any restriction against harassment of threatened or endangered species and defines prohibited harm to listed species as direct actions that actually kill or injure a member of a listed species.

The act allows the secretary to issue incidental take permits for acts necessary to establish and maintain experimental populations, necessary for research and implementation of captive propagation, or necessary for the public display or exhibition of wildlife for educational purposes. Before an incidental take permit can be issued, the applicant must develop a conservation plan specifying the impact on the species, steps the applicant can take to minimize those impacts, the funding available to implement those steps, and what alternative actions were considered and why they were rejected. The secretary cannot make the applicant include interests in land or water not owned by the applicant or address species other than the species for which the permit is sought.

In issuing take permits for scientific purposes, the secretary is required to give priority to research on alternative methods and technologies and their costs, to reduce incidental taking of species for which the use of existing methods and technologies entails significant costs to nonfederal persons.

The secretary is required to issue permits to any applicant who maintains a public display or exhibition of living wildlife and holds a license under the Animal Welfare Act, as well as to anyone who has demonstrated the ability to breed wildlife in captivity for release into the wild, maintenance of live specimens, or falconry. The act authorizes a permit holder to import, export, sell, purchase, or otherwise transfer possession of the affected species. Authorization of an activity under the Endangered Species Act, the Bald and Golden Eagle Protection Act, the Fish and Wildlife Conservation Act, the Lacey Act, the Marine Mammal Protection Act, the Migratory Bird Conservation Act, the Migratory Bird Treaty Act, or the Wild Bird Conservation Act will be consolidated into a general permit to cover all authorized activities. The act exempts captive-bred wildlife from the Endangered Species Act unless it is intentionally and permanently released to the wild.

H.R. 2275 prohibits the secretary from requiring nonfederal persons to provide any additional protection to aquatic species if the number of individuals of a listed species, including hatchery populations, exiting aquatic habitat under the control of the nonfederal person exceeds the number of individuals of the species entering the habitat under the nonfederal person's control.

The act limits the secretary's authority to act to conserve foreign species to actions that are consistent with the wildlife conservation programs of foreign nations, unless the secretary can demonstrate by substantial evidence that the foreign programs are inadequate. The secretary is prohibited from adopting more restrictive conservation measures than those specified in the Convention on International Trade in Endangered Species of Wild Fauna and Flora (CITES).

The secretary is required to grant an exemption from turtle excluder device (TED) requirements to any trawl fishers whenever the secretary determines that the contribution of such fishers to a sea turtle conservation program, including beach protection programs in foreign nations, exceeds the harm to sea turtles caused by the failure to use a sea turtle conservation program.

H.R. 2275 requires the secretary to solicit information about the status of a species from states and interested nonfederal persons and to take into account efforts by states, local governments, nonfederal persons, and private organizations to protect the species. The secretary is mandated to give greater weight to empirical data than to projections or extrapolations based on modeling and must consider captive-bred populations in determining whether a species should be listed. Foreign species listed under CITES cannot be listed unless the secretary determines that CITES does not provide adequate protection.

H.R. 2275 defines "best scientific and commercial data available" as factual information that has been subjected to peer review and, to the maximum extent feasible, verified by field testing. The act defines "distinct population of national interest" as a distinct population of a vertebrate species that is not otherwise listed in the United States, Canada, or Mexico but that has been designated as of national interest by Congress. The act defines "foreign species" as a species occurring outside the United States, not including marine species, species having a significant population in the United States, or migratory species that migrate through U.S. territory. The act defines "imminent threat to the existence of" a species as a significant likelihood that a species will be placed on an irreversible course to extinction within two years of listing unless the species is fully protected under the Endangered Species Act. The secretary of commerce is eliminated from the definition of "secretary" under the Endangered Species Act. H.R. 2275 defines "species" to include any subspecies and any distinct population of national interest. The secretary of the interior, upon proposing a species for listing, is required to publish a notice in the *Federal Register* for six months, requesting further information on the status of the species, subject to a six-month extension upon request of any person, and to give equal weight to any information submitted. The secretary cannot list a species on an emergency basis unless there is an imminent threat to the existence of the species. The secretary must publish any proposed listing in the *Federal Register* and provide a description of the data needed to ensure the scientific validity of the listing determination. He or she must also collect such data and consider it, with an opportunity for public review and comment. A peer review is required for any listing, delisting, critical habitat designation or revision, or jeopardy determination by reviewers appointed by the secretary and the governors of the states in which the species is located. Any petition to list a species must include an affidavit certifying that any scientific literature relied upon has been peer reviewed, listing the qualifications of any person asserting expertise on the species, and providing information on efforts to field-test population estimates.

The secretary must solicit information from the governor of each state in which a species proposed for listing is located, and if a governor advises against listing, the secretary must demonstrate that any information submitted by the governor is incorrect and that listing is warranted. The secretary must give at least 90 days notice before listing a species, through publication in the *Federal Register* and a newspaper of general circulation in each area where the species is believed to occur. The secretary must also give notice to and consult with the governor of each state and a representative of each county or local jurisdiction in which the species is believed to occur. If a state agency files written comments opposing the secretary's decision to list or not delist a species, the secretary is prohibited from carrying on unless the secretary consults with the president and submits a written justification to the state of his or her decision

H.R. 2275 allows any person to submit a petition for delisting, and the secretary is required to make a determination based on such a petition within 90 days.

The Endangered Species Act, Section 7, is modified to cause federal agencies to utilize their authorities to carry out programs for listed species conservation only when they are consistent with the agencies' primary missions. H.R. 2275 exempts from consultation requirement activities that are consistent with a conservation plan or, when no plan is required, a conservation objective for the species. An agency can, but is not required to, consult with the secretary if the agency judges that the activity is "likely to significantly and adversely affect" a listed species. If the acting agency determines that its responsibilities under the Endangered Species Act are in irreconcilable conflict with its responsibilities under other laws or treaties, it must request resolution by the president, who is to balance conservation with the public interest, economic and social factors, and the purposes of the conflicting law or treaty. A permit or license applicant may request that federal agencies consult with the secretary if the applicant believes his or her activities will affect a listed species, and the applicant must be allowed to be a full participant in the discussions. A federal agency must consult with the secretary if its activities might impact on critical habitat, causing destruction or adverse modification that is likely to jeopardize the continued existence of a species by significantly reducing the numbers or the distribution of the entire species. Unless a listed species is involved, federal agencies are prohibited from modifying any water allocation plan or any land management plan, standard, policy, or guideline. The secretary is required to provide a written statement of his or her opinion, including a summary of the information that the decision is based on and whether the agencies' plan is consistent with the conservation process. The consultation period is limited to 90 days, with a 45-day extension if agreed on by both parties, and if the deadline is not met, the acting agency's responsibilities are considered satisfied.

H.R. 2275 allows the U.S. Forest Service and the Bureau of Land Management to fund, authorize, and carry out actions consistent with land use and resource management plans under the National Forests Management Act and the Federal Land Management and Policy Act. This can be accomplished prior to completion of an Endangered Species Act Section 7 consultation if the agency believes the action "is not likely to significantly or adversely affect" listed species. Actions are exempt if they address an imminent threat to public safety or a natural catastrophe or to comply with federal, state, or local safety or public health requirements. Actions that are required for the routine operation, maintenance, and replacement of a federal or nonfederal project or facility are also exempt, as are actions that permit activity on private land.

H.R. 2275 would abolish the Endangered Species Committee, require the secretary to grant an exemption to the secretary of defense for reasons of national security, and allow the president to grant exemptions in major disaster areas for the repair or replacement of public facilities if they are necessary to prevent recurrence of the disaster and reduce the potential for loss of human life.

H.R. 2275 redesignates the Endangered Species Act's Section 5 as Section 5A and creates a new Section 5. The new section requires the secretary to form an assessment and planning team within 30 days after a listing. The team consists of public- and private-sector biologists, experts in property law and regulation, economists, and representatives of affected states and affected local governments. The team, within 180 days, must provide the secretary with a biological assessment that includes an assessment of the biological significance of the species, population trends, and minimum habitat needed to maintain the existence of the species and to recover the species. The assessment must also include a discussion of the technical practicality of recovering the species and describe potential management measures, including predator control, feeding, captive breeding, and experimental populations. Any demonstrable commercial or medicinal value of the species must be evaluated, and an economic assessment must be made of the impact on the public and private sectors, including local governments, that may result from the listing. The impact of potential management measures—including direct costs; impacts on tax revenue, employment, and use or value of property; and social, cultural, and community values—must also be assessed. The assessment must also include an evaluation of commercial activity that could potentially result in a net benefit for the species, as well as an intergovernmental evaluation of the impacts of the listing and potential management of species on state and local land use laws, conservation measures, and water allocation policies.

Within 210 days of listing, the secretary must evaluate the assessment and select a conservation objective from these options:

(1) recovery of the listed species, (2) a level of conservation that the secretary determines to be practicable and reasonable when weighed against the economic and social costs of implementation of conservation measures, (3) no federal action other than prohibition against direct, intentional harm to individuals of the species, and (4) another objective that the secretary determines will provide no less protection than the third objective. The secretary is required to publish the conservation objective in the *Federal Register,* but he or she is not required to submit determinations of conservation objectives for public review and comment. Within 12 months of listing, the secretary is required to publish a draft conservation plan, and within 18 months a final plan. The secretary is required to give priority to multispecies plans in geographic areas where there will be minimal conflict between conservation and economic activity, as well as plans that provide protection of species on units of the National Biological Diversity Reserve. The secretary must give priority to plans that have the least social and economic costs and plans that are implemented by states and private parties. The plan must describe what constitutes a prohibited take and provide recommendations on how to avoid a prohibited take. The plan must also contain alternate strategies and a description of the direct and indirect costs of each to the public and private sectors. It must also describe the effects on the use and value of property and on social, cultural, and community values. The plan must contain objective and measurable criteria for delisting. Each plan is required to provide fair and equitable treatment of states and nonfederal persons. The secretary is required to publish the conservation plan in the *Federal Register* and a newspaper of general circulation and to request public comments and provide appropriate public meetings.

H.R. 2275 authorizes federal agencies, at their own discretion, to enter into consultation with the secretary to determine if their actions are consistent with a conservation plan or objective and allows federal agencies to determine if their actions are consistent with the conservation plan or objective. The secretary is required to review each conservation plan and objective every five years and revise the plan if it is not meeting requirements, if the funding is not available to implement the plan, or if a more cost-effective measure is devised. The secretary must revise the plan if, on the basis of new data, he or she determines that the plan will not achieve the conservation objective.

H.R. 2275 redefines critical habitat as a specific area within a geographic area occupied by the species at the time of listing that contains features essential to the persistence of the species over a 50-year period and requires special management or protection.

The secretary must take into consideration economic and other relevant impacts of the designation and the listing on the affected area and must use the best scientific data available when designating critical habitat. The secretary must give priority to the National Biodiversity Reserve in designating critical habitat and must submit critical habitat proposals to the Bureau of Labor Statistics for comment. H.R. 2275 authorizes judicial review of critical habitat designations and authorizes the reviewing court to rule on the merits of the designation, rather than deferring to the secretary's expertise in designating critical habitat. The secretary must recognize to the maximum extent practicable captive breeding as a means of protecting and conserving listed species and provide grants to nonfederal persons for captive breeding programs. The secretary must determine whether the release of experimental populations is in the public interest and is essential to the continued existence of the species, as well as identify the precise boundaries of the geographic area for the release. However, members of experimental populations of a species designated as threatened must not be protected by the Endangered Species Act if they are found outside the geographic area of release and pose a threat to the welfare of the public. In the regulations the secretary must identify the measures taken to protect the public and domestic animals, identify the funding to pay for such measures, and implement the measures for experimental populations released outside the park or refuge systems, and the secretary must obtain written consent from the landowner for any release on nonfederal land.

Take resulting from routine operation, maintenance, and repair of buildings, other structures, roads, dams, airports, irrigation systems, or other facilities operated prior to the listing of the species shall be considered incidental take, and listed species are only protected from direct, intentional harm to individuals of the species.

H.R. 2275 requires that the secretary issue specific regulations for the protection of threatened species and mandates that these regulations be less restrictive than those for endangered species.

H.R. 2275 mandates the establishment of a National Biodiversity Reserve System on wholly owned federal land within the National Park System, the National Wildlife Refuge System, the National Wilderness Preservation System, and the National Wild and Scenic River System. The secretary can designate a unit of the reserve on state land if such a unit is nominated by the governor, and on private land if nominated by the owner, but is required to remove such lands from the reserve if the governor or the landowner request it. The reserve must be managed for biodiversity only to the extent that such management is consistent with the original purpose of the unit, with other laws applicable to the unit, and with activities that occur on it. The designation of a reserve unit cannot affect any existing permit, right, right-of-way, access, interest in land, right to use and receive water, or property right and requires that, within one year of designation, the manager complete an inventory of species on the reserve unit and make it available by public notice in the *Federal Register*. The Endangered Species Act shall not affect the water or water-related rights of the United States, the states, or any person or affect hunting, fishing, or other wildlife harvest laws, rules, and regulations under state law, federal law, or Indian treaty. The secretaries of the interior and agriculture are required to encourage exchanges of interests in land and waters within their jurisdictions for nonfederal lands and waters affected by the Endangered Species Act but may not impair existing easements, right-of-ways, water sources, and uses of adjacent land.

The secretary is authorized to delegate to a state the authority contained in the Endangered Species Act with respect to species of fish, wildlife, and plants that are resident in that state. The secretary may also determine that authority resides in the state agency to conserve listed species, that the state agency has established conservation programs consistent with the purposes of the Endangered Species Act for all resident listed species or such species as the state proposes to cover under its program, that the state agency is authorized to investigate species status, and that the state agency has established programs for habitat acquisition, conservation of listed fish and wildlife, public participation in listings, and voluntary or incentive-based conservation measures. The secretary can provide financial assistance to states to which Endangered Species Act authority has been delegated, in the form of money or real property.

H.R. 2275 authorizes funding through 2001 and mandates that the secretary pay 50 percent of the direct costs to nonfederal persons and federal power marketing agencies that were created to support the Endangered Species Act. H.R. 2275 also requires the secretary to pay 50 percent of the direct costs to the parties of a cooperative management agreement.

H.R. 2275 establishes an Endangered Species and Threatened Species Conservation Trust Fund made up of gifts, bequests, devises, and amounts appropriated to it. Amounts in the fund are to be available, subject to appropriations, for claims arising under the compensation provision, for habitat conservation grants, and for payment of federal cost sharing. H.R. 2275 has been referred to the Committee on Resources and the Committee on Agriculture. Committee hearings have been held, but there has been no more activity since 9 September 1996, when it was discharged by the House Committee on Agriculture and placed on Union Calendar No. 420.

Endangered Species Act Reform Amendments of 1995 (S.R. 768)

S.R. 768, sponsored by Senators Slade Gorton (R, Washington), J. Bennett Johnson (D, Louisiana), and Richard Shelby (R, Alabama) and introduced on 9 May 1995, would require the secretary of the interior to publish a notice of any proposed change in the status of a species and, upon the request of any interested person, appoint a three-person committee to review the scientific information and analyses for the proposed action and publish a report of the review panel. S.R. 768 would allow the protective effects of nonfederal regulatory mechanisms (that is, local, state, and international regulatory mechanisms, and non–Endangered Species Act federal mechanisms) to be considered when determining the threatened or endangered status of a species. S.R. 768 mandates that the secretary publish a description of the efforts to field-test the scientific data used to make the listing or conservation determination and that the secretary describe what additional information is needed to ensure the scientific validity of the determination, including a plan and deadlines for collecting the information. When making a listing proposal, the secretary must hold public meetings in each affected state. Prior to making a listing, the secretary must consider existing populations of the species in public and private captive breeding programs. S.R. 768 would allow judicial review of preliminary decisions to change the status of a species, and the bill changes the standard for emergency listing from "significant risk to the well-being" to "a significant likelihood that the species will be placed on an irreversible course to extinction during the 2-year period from the date of the listing determination unless the species is accorded fully the protections available under this Act during the period." The secretary would not be allowed to delegate this authority.

S.R. 768 relabels existing Endangered Species Act Section 5 as Section 5A and creates a new Section 5. The new section requires the secretary to form an "assessment and planning team" consisting of public- and private-sector biologists, land use specialists, economists, and representatives of affected state and local governments no later than 30 days after a listing determination. This team provides the secretary with a biological assessment of the species within 180 days after listing. This report must include an assessment of the biological significance of the species and the technical practicality of recovering the species based on population trends. It must also include potential management measures, including predator control, feeding, captive breeding, and experimental populations, and an economic assessment of the listing's potential impact on the public and private sectors, including direct costs and impacts on tax revenue, employment, and use or value of property. Within 30 days of receiving this report, the secretary must select a conservation objective for the species from the following options: (1) recovery of the listed species, (2) a level of conservation considered by the secretary to be "practicable and reasonable" when weighed against the "human and economic costs of implementation" of conservation measures, (3) no federal action other than prohibition against direct, intentional harm to individuals of the species, and (4) another objective that the secretary determines will provide no less protection than the third objective. The secretary can authorize a conservation objective greater than the third for distinct population segments only if it is in the "national interest based on biological, social, and economic factors."

The secretary is required to publish the draft plan, including critical habitat, within one year and the final plan and critical habitat within 18 months and must give priority to multispecies plans, as well as plans that have the lowest social and economic costs. Each plan must include alternative strategies, an analysis of the risks associated with each strategy, an estimate of the direct and indirect cost to the public and private sectors, and objective, measurable criteria for evaluating the strategy. *Critical habitat* is defined as an area within the geographic space occupied by the species that is essential for its continued existence over a 50-year time period. S.R. 768 requires the secretary to consult with the governor of each affected state, publish in the *Federal Register* and a newspaper of general circulation a summary of each plan, and hold at least two public meeting in each state. The secretary, at his or her discretion, designates critical habitat based on the best scientific data available and after consideration of the economic and other impacts and the benefits of designation. A copy of the plan must be submitted to the Bureau of Labor Statistics for written comment. The secretary is compelled to act only if failure to designate critical habitat will place the species on an irreversible path to extinction within two years.

S.R. 768 would alter Section 7 of the Endangered Species Act to allow federal agencies to determine whether their activities are consistent with a conservation plan or objective, without consultation with the secretary. Federal agencies are not required to conserve species prior to the publication of the final conservation plan.

If costs incurred by a nonfederal person under the conservation plan go unpaid, the measures that apply to that person will be suspended until the contribution is paid. Nonpayment for four years on any measure that restricts the title to property will eliminate the restriction.

The secretary is required to report to the congressional authorizing committee every two years about the status of efforts to develop and implement conservation plans and the status of species covered by existing plans. The secretary is also required to review conservation plans every five years and revise them as necessary. S.R. 768 provides for judicial review of the decisions of the secretary and any federal agency to determine if the decisions are in accordance with law. S.R. 768 amends the Marine Mammal Protection Act (MMPA) to substitute conservation plans for recovery plans.

After S.R. 768 is enacted, the secretary has 30 days to provide a list of all endangered and threatened species for which no final recovery plans have been issued. For listed species that occur in more than one state, the secretary must issue conservation objectives within 210 days, draft conservation plans within one year, and a final conservation plan within 18 months. For species limited to one state, the secretary is required to issue conservation objectives as expeditiously as possible. The recovery plan for any species is rescinded when a conservation objective has been published. The secretary is required to review biological opinions issued under Section 7 of the Endangered Species Act after 1 January 1995 and revise them in accordance with conservation objectives and plans.

S.R. 768 redefines *jeopardy* as actions or activities that significantly diminish the likelihood of the survival of the species in the wild by significantly reducing the number or distribution of the entire species.

S.R. 768 overrules the 1978 U.S. Supreme Court decision in *Tennessee Valley Authority v Hill,* which stated that Congress intended endangered species to be accorded the highest priority, and other duties of federal agencies must be reconciled with the conservation of endangered species. The act also eliminates consideration of cumulative impacts on listed species resulting from a modification of a public or private project. S.R. 768 prohibits any statement that emerges out of formal consultation from requiring, providing for, or recommending any restrictions or obligations on the activity of nonfederal persons if the activity is not authorized, funded, carried out, or otherwise subject to regulation by the federal government.

S.R. 768 abolishes the Endangered Species Committee, requires the secretary to grant exemptions to the Department of Defense in cases of national security, and allow the president to exempt areas declared disaster areas. The act also allows nonfederal persons to enter into consultation with the secretary, and if the secretary finds no jeopardy, he or she must provide a written statement to that effect. If the secretary finds that the actions do cause jeopardy, then he or she must suggest reasonable and prudent alternatives. The secretary may issue an incidental take permit, if it does not jeopardize the species.

S.R. 768 redefines *harm* within the definition of *take* as a direct action against any member of an endangered species of fish or wildlife that actually injures or kills a member of the species. The definition exempts take activities that are consistent with conservation plans or objectives or that address a critical, imminent threat to public health and safety or a catastrophic natural event. The act also exempts the incidental take of listed nonfish species within areas of the territorial seas and the exclusive economic zone not designated as critical habitat. The secretary is authorized to issue general permits, for a period of up to five years, for categories of activities that may affect listed species, if the secretary determines that the activities are similar in nature and will have only minimal separate and cumulative impacts on listed species. The general permits must be revoked if the secretary determines that the category of activities is having a greater than minimal adverse effect.

S.R. 768 requires the secretaries of the interior and agriculture to encourage exchanges of land and waters under their control for nonfederal land and waters affected by the Endangered Species Act. The value of the property must be treated as if the land or waters were not subject to Endangered Species Act restrictions.

The secretary is authorized to enter into cooperative agreements with state and local governments to manage areas within the control of these entities. Parties requesting agreements must perform the biological, economic, and intergovernmental assessments required for conservation objectives and plans. The secretary can rescind an agreement if, after a public hearing, he or she determines that the party is not administering or acting in accordance with the agreement. The secretary is required to pay 50 percent of the direct costs of implementing the agreement. Furthermore, the secretary is authorized to provide grants to private landowners to preserve habitat for listed species, if the property contains significant habitat and will be dedicated to habitat protection for a sufficient time to contribute to the protection of the species.

The secretary must make a determination of whether the release of experimental populations is in the public interest and must identify the precise boundaries of the geographic area for the release. If possible, the area of release should be within units of the National Park or National Wildlife Refuge systems. If releases occur outside these areas, the secretary must identify measures that protect the public and domestic animals, identify the funding to pay for such measures, and implement the measures. Experimental populations of threatened species must not be protected by the Endangered Species Act if they are found outside the geographic area of release and pose a threat to the welfare of the public. Furthermore, the secretary is authorized to provide grants to nonfederal persons for captive propagation programs for listed species.

S.R. 768 mandates that federal agencies are to implement the Endangered Species Act in a manner that ensures that people are not denied the reasonable use of their private property and that avoids any significant diminishment of the value of private property. S.R. 768 also mandates that such agencies are to balance achieving the conservation objective with ensuring economic growth and strong local and state tax bases. S.R. 768 requires that nothing in the Endangered Species Act diminish or impede the right of property owners to seek compensation for lost use value of property under the Fifth Amendment and other federal law and exempts private property of five contiguous acres or less, unless such activity poses an imminent threat to the existence of the species.

S.R. 768 allows any person, including any person who sustains actual or imminent economic injury as a direct or indirect result of a violation of the Endangered Species Act, to commence a civil suit to remedy violations of the Endangered Species Act or its accompanying regulations. S.R. 768 also allows any person to intervene as a matter of right in any suit brought under the act that threatens to cause injury to that person. Any water right acquired or used for any purpose under the Endangered Species Act shall be governed by the law of the state in which the water is used, including any priority system of water uses.

S.R. 768 requires the secretary to pay 50 percent of the direct and indirect costs, not to exceed $10 million, to nonfederal persons and federal power marketing agencies for complying with a conservation plan or the requirements under Section 7 of the Endangered Species Act. The secretary is authorized to acquire nonfederal land for habitat. He can exchange it for federal land, an interest in federal land (such as grazing rights), or appropriated funds. The secretary must use the pre–Endangered Species Act costs, and in the event of nonpayment, the requirements of the conservation plan or of Section 7 are suspended until full payment is made.

The secretary is authorized to issue educational permits for the public exhibition of listed wildlife designed to contribute to the education of the public about the ecological role and the conservation needs of the affected species, as long as the institution is licensed to do so under the Animal Welfare Act. Educational permits also authorize the permittee to import, export, buy, and sell the affected species. The terms *bred in captivity* and *captive bred* mean wildlife, including eggs, born or otherwise produced from parents in captivity. S.R. 768 reauthorizes the Endangered Species Act until 2001.

On 9 May 1995 this bill was read twice and referred to the Committee on Environment and Public Works. No legislative activity has occurred since then.

The Endangered Species
Reauthorization Act of 1997 (ESRA)

Section 2 of ESRA, a bill sponsored by Senators Dirk Kempthorne (R, Idaho), John Chafee (R, Rhode Island), and 14 other cosponsors, that was introduced on 16 September 1997, requires the secretary of the interior to use the best scientific or commercial data available and to give greater weight to data that is empirical, field-tested, or peer-reviewed. Petitions to list, delist, or change the listing status of a species should include information on the status and trends of all extant populations of the species, an appraisal of available data on threats to the species, and an indication of whether the information in the petition has been peer-reviewed or field-tested. Petitions to change the listing status of a species should be based on a change in the listing factors noted in Section 4(a)(1) of the Endangered Species Act, new data or a reinterpretation of existing data that indicates delisting is appropriate, extinction of the species, or the achievement of recovery goals. The act allows any person to petition for peer review of a proposed listing decision within 60 days of the proposal's publication in the *Federal Register*. The petitioner has to show "that there is substantial scientific or commercial information that is the basis for questioning the sufficiency or accuracy of the available data relied upon for the determination." A peer review panel is called from a list provided by the National Academy of Sciences. The panel consists of three independent "referees" who have demonstrated scientific expertise in the relevant subject area, do not have personal conflicts, and are not participants in other, ongoing listing processes.

The secretary is required to publish a description of additional information that would be helpful in the preparation of a recovery plan, including information that can and cannot be collected prior to the appointment of a recovery team. The standard for emergency listings would be changed from "posing a significant risk to the well-being" to "posing an imminent threat to the continued existence."

The secretary is required to notify appropriate state agencies and provide a period of at least 90 days in which the state can comment before publishing a listing proposal. He or she must also provide, upon request by any person, at least one public hearing on a proposed listing decision. The secretary must publish more detailed information than required by the Endangered Species Act on the status of the affected species, including "current population, population trends, current habitat, food sources, predators, breeding habits, captive breeding efforts, governmental and nongovernmental conservation efforts, or other pertinent information." The secretary must make the information used in listing decisions publicly available, with two types of exceptions: (1) the same exceptions that apply, for reasons of privacy and other reasons, to Freedom of Information Act requests (see 5 U.S.C. 552[b]) and (2) good cause, which may include a finding that disclosure could place the species at greater risk.

Section 3 of ESRA requires recovery plans except when the secretary finds that such a plan will not promote the conservation of a species because an existing plan or strategy to conserve the species already serves as the functional equivalent of a recovery plan. Recovery plans must offer alternative ways to achieve their goals. The alternatives must include an inventory of existing regulatory mechanisms for protecting the species; an estimate of the effectiveness of the alternatives; site-specific actions, research, captive breeding, establishment of refuges, or releases of experimental populations. Alternatives should use to the maximum extent practicable federal lands and federal agencies. Recovery plans should provide, where possible, opportunities to cooperate with state and local governments and other persons. They should also include a description of economic effects, including any significant impact on a particular geographic area or industry segment or on government revenues. A recovery plan should also include a description of ways an Endangered Species Act Section 10 conservation plan could help recover the species. Draft recovery plans should include a description of the scientific and commercial information on which the plan is based, as well as information that would be helpful in developing and implementing the plan and a strategy to obtain that information. Recovery plans should be cost-effective and efficient, and the secretary must give priority to plans that address significant and imminent threats to the survival of a listed species and have the greatest likelihood of achieving recovery. Whenever possible, plans should benefit species that are more taxonomically distinct and should cover multiple species, including candidate species. Plans should reduce adverse social and economic consequences to the maximum extent practicable and should reduce conflicts with construction or other development projects and other forms of economic activity. The plan should include goals that, when met, allow the species to be removed from the list. The goals should consist of objective, measurable scientific criteria as well as shorter-term benchmarks to help mark progress.

The secretary, in cooperation with affected states, appoints a recovery team that is broadly representative of constituencies with an interest in the species and its recovery and in the economic and social impacts of recovery. The recovery team may include representatives of federal agencies, tribal and local governments, and academic institutions; private individuals; organizations; commercial enterprises; and the secretary. Members are selected for their knowledge of the species or their expertise in the elements of the recovery plan or its implementation. The recovery team prepares a draft recovery plan, including advice on the designation of critical habitat. To improve public access to and participation in the recovery planning process, the secretary publishes a notice of availability of, a summary of, and a request for public comment on a draft recovery plan in a newspaper of general circulation in each affected state, as well as offering at least one public hearing in each affected state. The secretary, in making a habitat designation, must use only the best scientific and commercial data available, but the secretary is also required to consider and describe the economic impact of the designation. If the benefits of exclusion outweigh the benefits of designation, the secretary is authorized to exclude an area from critical habitat, unless the exclusion would result in the extinction of the species.

Section 4 of ESRA defines *reasonable and prudent alternatives* as alternative actions identified during consultation that can be taken with the intended purpose of the activity under consultation (for example, determining whether or not to build a dam in a specific location), that are economically and technologically feasible, and that can be implemented within the scope of the federal agency's legal authority. All federal agencies are authorized to implement recovery plans and are required to notify the secretary if they determine that an action may affect an endangered or threatened species, but not if the agency determines that the action is not likely to adversely affect that species. ESRA also allows multiagency consultations to be consolidated to a single agency. Applicants for incidental take permits are allowed to submit and discuss information with the secretary before the development of a draft biological opinion and to determine the information that the draft opinion is based on. If the secretary does not adopt a reasonable and prudent alternative, he or she is required to explain his or her failure to do so.

Section 5 of ESRA favors multispecies conservation plans and plans that minimize the impact on small landowners. State and local governments are responsible for enforcing conservation plans for each species. Low-effect conservation plans can be authorized; they provide an abbreviated and economically feasible permit process for any activity that will have minor or negligible effect on species. Safe harbor agreements allow a nonfederal landowner to create, restore, or improve occupied or unoccupied habitat for listed species. The secretary is authorized to provide cash grants of up to $10,000 to help develop safe harbors.

Section 6 of ESRA mandates that it be a policy of Congress that nothing in the Endangered Species Act will affect or limit the authority of a state to establish water rights or the use, control, appropriation, or distribution of water and that the federal government is subject to the procedural and substantive laws of the state, if the state were to alter these rights.

Under Section 7 of ESRA, a plaintiff in a citizen suit is required to establish that the acts of a person have (or will) proximately and foreseeably caused the take of an endangered or threatened species before he or she can initiate an enforcement action. ESRA provides "standing" to sue to those who are socially or economically impacted by the act. Section 8 requires the secretary to provide private landowners with information, to respond to requests for information or assistance, and to recognize exemplary efforts to conserve species on private land. Section 9 provides authorization through fiscal year 2003.

Section 10 of ESRA exempts from the National Environmental Policy Act recovery plan development and implementation, entry into implementation agreements for recovery plans, the development and approval of low-effect conservation plans, and the development and approval of other conservation plans and incidental take permits. Candidate species are those the secretary has sufficient information to list, but for which listing is precluded because of pending proposals to list species with higher priority.

On 16 September 1997 the bill was read twice and referred to the Committee on Environment and Public Works. Committee hearings were held later that month and it was ordered to be reported favorably with an amendment in the nature of a substitute. On 31 October 1997 Senator Chafee gave the report to the Senate with written report No. 105-128. Additional and minority views were filed and it was placed on the Senate Legislative Calendar. No legislative activity has occurred since then.

Related Laws

The Bald Eagle Protection Act of 1940 (54 Stat 250, see 16 U.S.C. 668)4

The enacting clause of this act provided the following:

> Whereas the Continental Congress in 1782 adopted the bald eagle as the national symbol; and Whereas the bald eagle thus became the symbolic representation of a new nation under a new government in a

new world; and Whereas by that act of Congress and by tradition and custom during the life of this Nation, the bald eagle is no longer a mere bird of biological interest but a symbol of the American ideals of freedom; and Whereas the bald eagle is not threatened with extinction: therefore Be it enacted . . .

The act (as amended by PL 86-70 [73 Stat 143], PL 87-884 [76 Stat 1246], and PL 92-535 [86 Stat 1064] made it unlawful for anyone in the United States or any place subject to its jurisdiction to knowingly or with wanton disregard for the consequences of his or her act take; possess; sell; purchase; barter; offer to sell, purchase or barter; transport; export; or import, at any time or in any manner, any bald eagle or golden eagle, alive or dead, or any part, nest, or egg thereof. The act defined civil and criminal penalties. The secretary of the interior can permit taking, possession, and transportation of specimens of bald or golden eagles for the scientific or exhibition purposes of public museums, scientific societies, and zoological parks or for the religious purposes of Indian tribes. The secretary may also authorize the taking of eagles for the protection of wildlife or agricultural or other interests.

National Wildlife Refuge System, Public Law 89–669 (80 Stat 927, see 16 U.S.C. 668dd)

This act requires the secretary of the interior to consolidate different categories of areas, such as wildlife ranges, game ranges, wildlife management areas, and waterfowl production areas, into the National Wildlife Refuge System. The act mandates that no one shall knowingly disturb, injure, cut, burn, remove, destroy, or possess any real or personal property of the United States, including natural growth, in any area of the system or take or possess any fish, bird, mammal, or other wild vertebrate or invertebrate animal or part, nest, or egg thereof within such an area. The secretary is authorized to promulgate regulations to control fishing and hunting, public recreation, and accommodations. The act provides civil and criminal penalties.

Migratory Bird Treaty Act (40 Stat 755, see 16 U.S.C. 703)

This act makes it unlawful at any time, by any means, or in any manner to pursue; hunt; take; capture; kill; attempt to take, capture, or kill; possess; offer for sale; sell; offer to barter; barter; offer to purchase; purchase; deliver for shipment; ship; export; import; cause to be shipped, exported, or imported; deliver for transportation; transport or cause to be transported; carry or cause to be carried; or receive for shipment, transportation, carriage, or export any migratory bird or any part, nest, or egg of any such bird, or any product, whether or not manufactured, that consists, or is composed in whole or in part, of any such bird or any part, nest, or egg thereof. All birds covered by conventions between the United States and Great Britain (16 August 1916), Mexico (7 February 1936), Japan (4 March 1972), and the Union of Soviet Socialist Republics (19 November 1976) are included under this act. The secretaries of agriculture and the interior may promulgate regulations that allow the taking of birds. The act provides civil and criminal penalties.

Migratory Bird Conservation Commission Act (45 Stat 1222, see 16 U.S.C. 715)

This act establishes the Migratory Bird Conservation Commission, which consists of the secretaries of agriculture and the interior, the administrator of the Environmental Protection Agency, and two members each of the Senate and the House of Representatives. The commission is authorized to consider and pass upon the purchase or rental of any area of land, water, or land and water recommended by the secretary of the interior that is required for the conservation of migratory birds.

Wild Free-Roaming Horses and Burros Act, Public Law 92–195 (85 Stat 649, see 16 U.S.C. 1331)

Congress found that wild and free-roaming horses and burros are a living symbol of the historic and pioneer spirit of the West, contribute to the diversity of life forms within the nation, and enrich the lives of the American people. Since these horses and burros are disappearing from the American scene, it is a policy of Congress that free-roaming horses and burros be protected from capture, branding, harassment, or killing. Furthermore, they are to be considered an integral part of the natural system of the public lands on which they are found. Free-roaming horses and burros are placed under the jurisdiction of the secretary who is directed to protect and manage them as components of the public lands. He or she is to designate and maintain specific ranges on public lands as sanctuaries for the animals' protection and preservation, based on consultations with qualified scientists. The secretary is directed to maintain an inventory of animals and to determine the appropriate management levels of animals. If overpopulation exists, he or she determines what measures should be undertaken. The secretary is to destroy old, sick, and lame animals in a humane manner and is to provide for the human capture of excess animals, to be placed for private maintenance. The act provides for criminal and civil penalties.

Marine Mammal Protection Act, Public Law 92–522 (86 Stat 1027, see 16 U.S.C. 1361)

Congress found that certain species of marine mammals were or might be in danger of extinction or depletion as a result of the activities of humans. At the effective date of the act, 21 October 1972, a moratorium on the taking and importing of marine mammals and marine mammal products began, except in cases in which permits are issued by the secretary. One immediate goal was to reduce the incidental killing or serious injury of marine mammals as a result of commercial fishing to levels approaching zero. The secretary is directed to ban the importation of commercial fish or products made of fish caught by commercial fishing technology that results in the incidental kill or serious injury of marine mammals. The secretary is directed to use the best scientific information available and to consult with the Marine Mammal Commission to determine when and to what extent, if any, it is appropriate to allow the taking of marine mammals. The act permits Indian, Aleut, and Eskimo individuals who reside in Alaska on the coast of the Pacific Ocean or the Arctic Ocean to take marine mammals for subsistence purposes or for the purpose of creating and selling authentic native articles of handicrafts and clothing, if the taking is accomplished in a nonwasteful manner. The secretary may grant certain hardship exemptions. The act establishes civil and criminal penalties.

Marine Mammal Commission Act, Public Law 92–522 (86 Stat 1043, see 16 U.S.C. 1401)

This act directs the president to appoint, with the advice and consent of the Senate, a commission consisting of three members, drawn from a list provided by the chair of the Council on Environmental Quality, the secretary of the Smithsonian Institution, the director of the National Science Foundation, and the chair of the National Academy of Science. Commissioners must be knowledgeable about marine ecology and resource management and must not be in a position to profit from the taking of marine mammals. The commission is to review and study the activities of the United States pursuant to existing laws and international conventions relating to marine mammals. This includes, but is not limited to, the International Convention for the Regulation of Whaling, the Whaling Convention Act of 1949, the Interim Convention on the Conservation of North Pacific Fur Seals, and the Fur Seal Act of 1946. The commission reviews the condition of stocks of marine mammals, methods used to protect and conserve marine mammals, and research programs to study marine mammals. The commission consults with the secretary on matters relating to marine mammals.

The National Environmental Policy Act of 1969, Public Law 91–190 (83 Stat 852, see 42 U.S.C. 4321)

This act establishes a Council on Environmental Quality and sets forth a national policy that encourages productive and enjoyable harmony between humans and the environment and that promotes efforts to prevent or eliminate damage to the environment and the biosphere. The act encourages efforts that enrich the understanding of the ecological systems and natural resources important to the nation. Congress recognizes the profound impact of human activity, especially population growth, high-density urbanization, and industrial expansion, on all components of the natural environment. The act holds that it is vitally important to restore and maintain environmental quality and that it is a responsibility of the federal government to assure for all Americans safe, healthful, productive, and aesthetically and culturally pleasing surroundings. The act mandates that the federal government preserve important historic, cultural, and natural aspects of our national heritage and attain the widest range of beneficial uses of the environment, without degradation or risk to health and safety. The federal government is directed to utilize a systematic and interdisciplinary approach and to use the natural and social sciences, as well as the environmental design arts, in planning and decision making that impact our environment. Any major federal action that might have an impact on the quality of the environment must include a detailed statement about the environmental impact of the proposed action, a description of any adverse environmental effects that cannot be avoided should the proposal be implemented, alternatives to the proposed action, a statement of the relationship between local short-term uses of the environment and the maintenance and enhancement of long-term productivity, and any irreversible and irretrievable commitments of resources that would be involved in the proposed action should it be implemented.

The act mandates that the president transmit to Congress an Environmental Quality Report that describes the status and condition of major natural, man-made, and altered environments in the United States, including the air, water, and terrestrial environments. This report must also describe current and foreseeable trends in the quality, management, and utilization of these environments and the effects of those trends on the social, economic, and other requirements of the nation, as well as the adequacy of the available natural resources for fulfilling the human and economic requirements of the nation, especially in light of population pressures. The report should also review the programs and activities of the federal, state, and local governments with respect to their effects on the environment and on the conservation, development, and utilization of natural resources and should describe a program for remedying any deficiencies in existing programs and activities.

The act creates the Council on Environmental Quality, which consists of three members who are appointed by the president and who serve at his or her pleasure, with the advice and consent of the Senate. Each member, by training, experience, and attainment, should be qualified to analyze and interpret environmental trends and information, to appraise programs and activities of the federal government, to be aware of and responsive to scientific, economic, social, aesthetic, and cultural needs, and to formulate and recommend national policies to improve the quality of the environment.

Selected State Legislation

California: Fish and Game Code, Sections 2050 to 2068, 2070 to 2079, and 2090

California enacts the Endangered Species Act. The enacting clauses maintain that certain species of fish, wildlife, and plants have been rendered extinct as a result of human activities, untempered by adequate concern and conservation; that other species are in danger of or threatened with extinction because their habitats are threatened; that these species are of ecological, educational, historical, recreational, aesthetic, economic, or scientific value; and that the conservation, protection, and enhancement of these species and their habitat is of statewide concern. The legislature finds and declares that it is the policy of the state to conserve, protect, restore, and enhance any endangered or threatened species and their habitat and that it is the intent of the legislature to acquire lands for habitat. The legislature holds that state agencies should not approve projects that jeopardize the continued existence of these species, if reasonable and prudent alternatives can be developed. If alternatives are not available, projects are approved if they provide appropriate mitigation and enhancement. All state agencies, boards, and commissions must seek to preserve these species. The policy of the legislature is to foster cooperation with private landowners to protect these species.

Sections 2070 to 2079 establish a commission that lists endangered or threatened species, and adding and removing species based on scientific information. The commission accepts petitions from interested persons to add or remove species. Section 2090 mandates that state agencies consult with the Fish and Game Department to determine if any action authorized, funded, or carried out by the agency is likely to jeopardize the continued existence of a threatened or endangered species.

Texas: Parks and Wildlife Code, Chapters 67, 68, and 88

In 1973, the Texas legislature authorized the Texas Parks and Wildlife Department to establish a list of threatened and endangered animals. Laws and regulations dealing with threatened and endangered animal species are contained in Chapters 67 and 68 of the Texas Parks and Wildlife Code as well as Sections 65.171 to

65.184 of Title 31 of the Texas Administrative Code. In 1988, the Texas legislature established a list of threatened and endangered plant species, and the laws and regulations dealing with them are contained in Chapter 88 of the Texas Parks and Wildlife Code as well as Sections 69.01 to 69.14 of the Texas Administrative Code. The Resource Protection Division is responsible for the listing and recovery of threatened and endangered species. The department's legal division issues permits for the handling of listed species. The regulations prohibit the taking, possession, transportation, or sale of any of the listed species without a permit. The code also forbids collection of listed species from public lands. In 1983, the Texas legislature created the Special Nongame and Endangered Species Conservation Fund, used for nongame wildlife and endangered species research and conservation, habitat acquisition and development, and dissemination of information pertaining to these species. Money for the fund is obtained from private donations and the sale of nongame wildlife art prints, decals, and stamps.

Information about Federal Laws

Print Resources

U.S. Code (U.S.C.). Washington, DC: Government Printing Office

The official multivolume compilation of U.S. statutes. Two volumes of the U.S.C. are especially useful: the general index, which lists keywords that can be used to find statutes dealing with specific topics, such as animals, and the index of popular names of specific statutes. By using the latter index, it is possible to determine, for example, that the Endangered Species Act of 1973 is Public Law 93–205. Knowing a statute's number helps locate the text of the statute, its legislative history, and so on. This index also provides citations to Statutes at Large, a multivolume chronological compilation of laws as they are enacted. The Endangered Species Act is cited as 87 Stat 884.

United States Code Annotated (U.S.C.A.). St. Paul, MN: West Publishing Co.

An unofficial multivolume compilation of federal statutes and their amendments, arranged by topic. In addition to the text of each statute, the U.S.C.A. presents historical footnotes, annotations to law review articles, and cases dealing with statutory provisions. Updated annually in the form of pocket pieces and paperbound advance sheets.

U.S. Code Congressional and Administrative News, St. Paul, MN: West Publishing Co.

An unofficial compilation, arranged by congressional session, of federal statutes. The text of each statute is presented, along with its legislative history. The history includes an opening statement, a statement of the purpose of the statute, a summary of the outcome of congressional hearings, and the committees that worked on the statute and their reports. A chronological record of the actions of the House and the Senate with regard to the bill and the passage of the statute are included.

Code of Federal Regulations. Washington, DC: Office of the Federal Register, 1992

An official multivolume compilation of regulations and their amendments as issued by federal agencies, arranged by topic. Federal regulations are published chronologically in the *Federal Register*, then codified in the *Code of Federal Regulations*. Federal regulations dealing with endangered species are found in Title 50, Wildlife and Fisheries; Chapter IV, Joint Regulations (U.S. Fish and Wildlife Service, Department of the Interior and the National Marine Fisheries Service, National Oceanic and Atmospheric Administration, Department of Commerce); and Endangered Species Committee Regulations.

Internet Resources

The World Wide Web has a number of sites that can provide up-to-the-minute information about the status of pending bills, acts, and laws.

Thomas Legislative Information on the Internet (http://thomas.loc.gov) provides information about current bills and acts, as well as committee reports and the text of the *Congressional Record*.

The U.S. House of Representatives Internet Law Library (http://law.house.gov/2.htm) provides searchable online versions of the *U.S. Code* and the *Code of Federal Regulations*, as well as links to other resources.

International Treaties and Conventions

A convention is a pact or an agreement between states or nations in the nature of a treaty. It can take the form of agreements or arrangements preliminary to a formal treaty, which serve as its basis, or of international agreements for the regulation of matters of mutual concern, but coming within the sphere of politics or commercial intercourse. A treaty is a compact (a working agreement between or among states concerning matters of mutual concern) made between two or more independent nations with a view to the public welfare. A treaty is generally signed by commissioners that are properly authorized, and a treaty is solemnly ratified by the supreme power of each state. (The preceding discussion is based on *Black's Law Dictionary* [St. Paul, MN: West Publishing, 1979].) Multilateral treaties frequently contain provisions allowing the parties to release themselves from certain aspects of the treaties, if they wish. This is generally known as making a "reservation."

Treaties made for and on behalf of the United States are made by the president, with the advice and consent of the Senate (see the U.S. Constitution, Article II, Section 2). Specifically, after the president has negotiated and signed a treaty, it is submitted to the Senate for its advice and consent. The text of the treaty, as well as the presidential message and other documents, are published in the *Senate Executive Documents*. The treaty is assigned to the Committee on Foreign Relations. The committee may hold hearings and ultimately issues a report recommending that the treaty either be ratified or not be ratified. If a treaty is ratified, it is published in pamphlet form in the *Treaties and Other International Acts Series* (T.A.I.S.) and published in bound volume in *United States Treaties and Other International Agreements* (U.S.T.).

One of the first international treaties to protect wildlife was the Treaty Concerning the Regulation of Salmon Fishing in the Rhine River Basin, signed by Germany, Luxembourg, the Netherlands, and Switzerland in 1886. Animals that live in territory that belongs to nobody, such as Antarctica and the high seas, are especially vulnerable, as are migratory species. The International Union for the Conservation of Nature and Natural Resources (IUCN) was formed in 1948 and includes among its members governments, governmental agencies, and nongovernmental organizations from 111 countries.

Convention Relative to the Preservation
of Fauna and Flora in Their Natural State

On 14 January 1936, this convention proclaimed that the fauna and flora of certain parts of the world, especially Africa, were threatened with permanent injury or extinction. Preservation of these species was agreed to be best achieved by developing national parks and nature and other reserves where the hunting, killing, and capturing of fauna and the collection and destruction of flora is limited or prohibited. In nature reserves, according to the convention, it is strictly forbidden to engage in any form of hunting or fishing; any undertaking connected with forestry, agriculture, or mining; any excavating, prospecting, drilling, leveling of the ground, or construction; any work that alters the configuration of the soil or the character of the vegetation; any act likely to harm or disturb the fauna and flora; or the introduction of any species of fauna and flora, whether indigenous or imported, wild or domesticated. No one is allowed to enter, traverse, or camp in a reserve without a written permit from competent authorities. Scientific investigations may only be undertaken by permission of those authorities. The authorities must establish an intermediate zone around the borders of a national park or a nature reserve where the hunting, killing, and capturing of animals is controlled by the authorities of the park.

The convention provides lists of animals that should be protected as completely as possible (Class A) and animals that do not require as rigorous protection (Class B). The convention limits the rights of indigenous peoples. The convention does not preclude hunting or killing of protected species in times of famine; to protect human life, public health, or domestic stock; or to maintain public order.

The convention prohibits the use of motor vehicles and aircraft (including lighter-than-air aircraft) for hunting, killing, or capturing animals or to drive, stampede, or disturb them, including for the purpose of filming or photographing them. The convention prohibits the use of fire for hunting, capturing, or destroying animals. The use of poison or explosives for killing fish; the use of dazzling lights, flares, poisons, or poisoned weapons for hunting animals; and the use of nets, pits, enclosures, gins, traps, or snares, as well as set guns and missiles containing explosives, are prohibited.

Convention between the United States of America and Mexico for the Protection of Migratory Birds and Game Mammals

Proclaimed 15 March 1937, this convention allows both countries to determine how migratory birds can be used rationally for sport, food, commerce, and industry. The convention mandates the establishment of closed seasons, when the taking of migratory birds, their nests, or their eggs is prohibited. The convention also mandates the establishment of refuge zones where the taking of these birds is prohibited. The closed season for wild ducks is from 10 March to 1 September. The convention prohibits the killing of migratory insectivorous birds, except when they become injurious to agriculture. The convention prohibits all hunting from aircraft. The convention prohibits the transport of migratory birds, dead or alive, or their parts or products over the U.S.-Mexican border without a permit issued by both countries. Article 4 lists the families of game and nongame migratory birds that are covered by the convention. Migratory game mammals are also covered by the convention. The convention was amended on 10 March 1972 by adding additional families of birds.

Convention on Nature Protection and Wildlife Preservation in the Western Hemisphere

Proclaimed 1 May 1942, this convention mandates the formation of national parks, reserves, and monuments to preserve areas of superlative scenery, as well as flora and fauna of national significance. Hunting, killing, or capturing fauna and collecting flora are prohibited within these areas except under the direction and control of the authorities and for duly authorized scientific investigations. Facilities for public recreation and education are provided within these areas. The convention mandates that the parties make suitable laws and regulations to protect and preserve flora and fauna that are not included in these areas, as well as to enact laws that assure the protection and preservation of the natural scenery, striking geologic formations, and regions and natural objects of aesthetic interest or of historic or scientific value. The parties also agree to provide protection for migratory birds of economic or aesthetic value and to prevent the extinction of any species. The convention provides a list of species that must be afforded special protection.

International Convention for the Regulation of Whaling

Proclaimed 26 November 1945, this convention applies to all factory ships, land stations, and whale catchers, which include any ships used for the purpose of hunting, taking, towing, holding onto, or scouting for whales. The convention establishes an International Whaling Commission consisting of one member from each contracting government. The commission elects a chair and a vice-chair, and a simple majority makes decisions. The commission encourages, recommends, and organizes investigations relating to whales and whaling; collects and analyzes statistical information concerning the current conditions and trends of whale stocks and the effects of whaling activities; and studies, appraises, and disseminates information about the methods of maintaining and increasing populations of whales. The commission may amend from time to time the provisions of the hunting schedule by adopting regulations with respect to the conservation and utilization of whale resources, fixing (1) protected and unprotected species, (2) open and closed seasons, (3) open and closed waters, including sanctuary areas, (4) size limits for each species, (5) time, methods, and intensity of whaling (including the maximum catch of whales to be taken in any one season), (6) types and specifications of gear, apparatus, and appliances that may be used, (7) methods of measurement, and (8) catch returns and other statistical and biological records. Changes in the schedule must be based on scientific findings and must not involve restrictions on the number or nationality of factory ships or land stations or allocate specific quotas to any factory ship or land station or to any group of factory ships or land stations. The commission must take into consideration the interests of the consumers of whale products and the whaling industry in making its decisions. The convention provides that any contracting government may grant to any of its nationals a special permit authorizing that national to kill, take, and treat whales for purposes of scientific research.

International Convention for the Protection of Birds

The purpose of this convention, proclaimed 17 January 1963, is to protect birds in the wild state. Protection is granted to all birds during their breeding seasons and during their return flights to nesting grounds. Protection is granted throughout the year to all birds that are in danger of extinction or that are of scientific interest. The import, export, transport, sale, offer for sale, purchase, giving, or possession of any live or dead bird or any part of a bird during the season when the species is protected is prohibited. The removal or destruction of nests or the taking, damaging, transport, import, export, sale, offer for sale, purchase, or destruction of eggs or their shells or broods of young birds in the wild state during the season when they are protected and particularly during their breeding season is prohibited. Methods used to enumerate birds that result in mass kill, the capture of birds, or the unnecessary suffering of birds are prohibited. The parties must gradually introduce into their legislation measures designed to prohibit or restrict the use of snares; bird lines; traps; hooks; nets; poisoned bait; stupefying agents; blinded decoy birds; decoy ponds with nets; mirrors, torches, and other artificial lights; fishing nets or tackle for the capture of aquatic birds; magazine or automatic sporting-guns holding more than two cartridges; all firearms other than shoulder arms; and the pursuit and shooting of birds from motorboats in inland waters. Also prohibited should be the use of motor vehicles or airborne machines to shoot or drive birds, the offering of rewards for the capture or killing of birds, and all other methods designed for the mass capture or killing of birds. If a species is found to damage fields, vineyards, gardens, orchards, woods, game, or fish or threatens to destroy or diminish one or more species whose conservation is desirable, authorities may issue permits to take the former species. The convention mandates that the parties consider and adopt measures to prevent the destruction of birds by hydrocarbon emissions (for example, oil spills), water pollution, lighthouses, electric cables, insecticides, poisons, or any other means. The convention also mandates that the parties encourage and promote the creation of water and land reserves where birds can nest and raise their broods and where migratory birds can find rest and food undisturbed.

Convention on International Trade in Endangered Species of Wild Fauna and Flora (CITES)

Proclaimed 12 May 1975, this convention recognizes that wild fauna and flora are an irreplaceable part of the natural systems of Earth and that these fauna and flora are of value for aesthetic, scientific, cultural, recreational, and economic reasons. International cooperation is essential to protect certain species from overexploitation through international trade. The species listed in Appendix I of the convention are threatened with extinction, and trade in living or dead specimens of these species must be controlled. All species listed in Appendix II are not necessarily threatened with extinction but may become so unless trade is subject to strict regulations. All trade in species listed in these appendixes requires the prior grant and presentation of an export or import permit. The permit must not be granted unless it can be demonstrated that such export will not be detrimental to the survival of the species and that the taking of the species is in accordance with the laws of the country where the taking occurs. All living specimens must be prepared and shipped to minimize the risk of injury, death, or damage to the health of the animal and to prevent cruel treatment. The recipient of a living specimen must be suitably equipped to house and care for it, and the specimen must not be used primarily for commercial purposes. The convention proposes that each party penalize trade in or possession of specimens covered by the convention and provide for confiscation or return to the state of export any illegally taken specimens of species covered by the convention.

A secretariat is provided by the executive director of the United Nations Environmental Programme to help arrange and service meetings of the parties and to undertake scientific and technical studies in accordance with programs authorized by the conference, especially with regard to the shipment of living specimens and to their identification. The secretariat also publishes current editions of the appendixes and provide yearly reports to the parties on the work of the secretariat. Article XV provides a mechanism to propose amendments to Appendixes I and II for listing new species.

Convention on Biological Diversity

This convention mandates the conservation of biological diversity and the fair and equitable sharing of the benefits of utilizing genetic resources. Each party must develop national strategies and programs for the conservation and sustainable use of biological diversity. The parties must identify components of biological diversity that are important to conserve, identify processes and categories of activities that are likely to have adverse impacts, and monitor the effects of those activities through sampling and other techniques. Each party must provide and manage a system of protected areas where special measures are taken to conserve biological diversity, where ecosystems are to be protected, and where viable populations of species are to be maintained in natural surroundings. Environmentally sound development is to be promoted in areas adjacent to the protected areas. Parties must rehabilitate and restore degraded ecosystems and promote the recovery of threatened species. The parties must also establish the means to regulate the risk associated with the release of organisms modified by biotechnology, and parties must either prevent the introduction of, control, or eradicate alien species that threaten ecosystems, habitats, or species. Participating states should respect, preserve, and maintain the knowledge, innovations, and practices of indigenous and local communities embodying traditional lifestyles relevant to the conservation and sustainable use of biological diversity. Those countries should also establish and maintain facilities for the ex situ conservation of and research on plants, animals, and microorganisms, preferably in the country of origin of genetic resources. Programs should be established to train scientific and technical personnel in the identification, conservation, and sustainable use of biological diversity. The conservation and sustainable use of biological diversity should be promoted in developing countries. The convention mandates that parties provide or facilitate access and transfer to other parties of technologies that are relevant to the conservation and sustainable use of biological diversity or that make use of genetic resources and do not cause significant damage to the environment.

U.S. Endangered Species Litigation

5

Sources of the Law

The American legal system is based on the common-law tradition that originated and developed in England. It consists of rules and principles of action that are applicable to government and the security of persons and property that do not rest for their authority on the positive declarations of the will of the legislature. Common law is sometimes referred to as unwritten law.

Early in the history of English law, the decisions of courts came to be considered precedents. That is, once a court has laid down a principle of law as applicable to a certain set of facts, it adheres to that principle and applies it to all future cases in which the facts are the same. This doctrine is commonly called *stare decisis*. The federal system and each of the 50 states have a hierarchy of courts that adhere to this doctrine.

The Endangered Species Act of 1973 and its amendments have generated considerable litigation in both the state and federal courts. The selected cases reviewed in the first section of this chapter are designed to give the flavor of the case law as applied to the act. A good deal of the case law has focused on the status of the northern spotted owl *(Strix occidentalis caurina)*, and selected cases are reviewed in the second section of this chapter (see also Chapter 1).

It is likely that if the Endangered Species Act of 1973 is reauthorized in its present form and seeks protection of threatened or endangered species on nonfederal (that is, private) lands, the amount of litigation will increase significantly. If any of the bills presently before Congress that specifically seek to protect threatened or endangered species on nonfederal land are enacted, there will be a significant increase in litigation. Private citizens would undoubtedly seek relief under the Fourth Amendment to the United States Constitution, which guarantees people the right to be secure in their homes and property against unreasonable seizures.

General Endangered Species Act Cases

U.S. Supreme Court

Geer v. Connecticut, *161 U.S. 519*

The issue in this case is whether it is lawful under the Constitution of the United States (Article I, Section 8) for the state of Connecticut to allow the killing of birds within the state during a designated open season; to allow such birds, when so killed, to be used, sold, or bought for use within the state; and yet to forbid their transportation outside the state. This is the first case the Supreme Court decided relating to wild animals.

From the earliest traditions, the right to reduce animals from *ferae naturae* to possession has been under the control of the lawgiving power. Roman common law embraced the concept of *ferae naturae,* under which animals having no owner were considered to belong in common to all the citizens of the state. Natural law holds that wild animals belong to whoever takes them, but civil law can restrict what natural law permits. So sovereigns could reserve the right to hunt all game for themselves and those they chose to transmit it to and forbid all others to hunt. The common law of England based property in game upon the principle of common ownership and therefore treated it as subject to governmental authority. The common-law right of government to control the taking of animals was vested in the American colonial governments and ultimately in the states with the separation from the mother country. This right remains with the states till this day, insofar as its exercise is not incompatible with or restrained by the rights conveyed to the federal government by the Constitution. Most of the states have passed laws for the protection and preservation of game. Many adjudicated cases recognize the right of the states to control and regulate the common property in game.

The Supreme Court of Minnesota has stated the principle as follows: "We take it to be the correct doctrine in this country, that the ownership of wild animals, so far as they are capable of ownership, is in the State, not as a proprietor, but in its sovereign capacity as the representative and for the benefit of all its people in common." It has never been judicially denied that the government may make regulations for the preservation of game and fish. The states have restricted the taking and molestation of wild animals to certain seasons of the year. The right to hunt and kill game is a boon or a privilege, granted expressly or implicitly by sovereign authority. It is not a right inherent in each individual.

In *Geer v. Connecticut*, the Court held that the state did have the authority to deny transportation of game, rightfully killed, outside the state.

Tennessee Valley Authority
v. Hiram G. Hill et al., *98 Sup. Ct. 2279*

Environmental groups and others brought action under the Endangered Species Act of 1973 to enjoin the Tennessee Valley Authority from completing the Tellico Dam and impounding a section of the Little Tennessee River (see *Hill et al. v. Tennessee Valley Authority* later in this chapter). The project began in 1967 as a multipurpose regional developmental project to provide electric current, a flat-water recreational area, and a flood control area. In 1973, a previously unknown species of perch, the snail darter (*Percina [Imostoma] tanasi*), a 3-inch, tannish fish was discovered near the Little Tennessee River. More than 130 known species of darters exist, and it is often difficult to differentiate one species from another. The snail darter was listed by the secretary of the interior as an endangered species on 8 October 1975. In making the listing, the secretary noted that "the snail darter is a living entity, which is genetically distinct and reproductively isolated from other fishes." The known population of snail darters is 10,000 to 15,000 individuals. Critical habitat for the species is swift shoals over clean gravel in cool, low-turbidity water. The snail darter's food consists of snails that live in these shoals. The critical habitat, described in the *Federal Register*, volume 41, pages 13926 to 13928 (see also 50 CFR 17.81), would be destroyed by the operation of the dam.

Respondents filed suit in district court under Section 11(g) of the Endangered Species Act, which allows any person to commence a civil suit to enjoin the United States or any other governmental instrumentality or agency that is alleged to be in violation of any provision of the act. The district court found that completing the dam would probably jeopardize the snail darter's continued existence (see 419 F. Supp. 753). However, the court found that Congress, aware of the snail darter problem, continued to fund the project: "At some point in time a federal project becomes so near completion and so incapable of modification that a court of equity (that is, a court where justice is administered according to fairness, rather than with the strictly formulated rules of evidence and precedent) should not apply a statute enacted long after inception of the project." The court of appeals (see *Hill et al. v. Tennessee Valley Authority* later in this chapter) reversed the district court and ordered it to enjoin the project "until Congress, by appropriate legislation, exempts Tellico from compliance with the Act or the snail darter has been deleted from the list of endangered species or its critical habitat materially redefined."

On further appeal, the Supreme Court held that completion and operation of the dam would either eradicate a known population of the snail darter, an endangered species, or destroy its critical habitat and thus prohibited completion of the dam. The Court did so even though the dam was virtually completed and Congress continued to appropriate large sums of public money for the project. The Court held that Section 7 of the Endangered Species Act is not limited to projects in the planning stages and that Congress intended to halt and reverse the trend toward species extinction whatever the cost. The language of the Endangered Species Act reveals a conscious decision to give endangered species priority over the primary missions of federal agencies and clearly shows that Congress viewed the value of an endangered species as incalculable. Continued appropriations for the project after the congressional appropriations committees were apprised of the effect of the project on the snail darter did not constitute implied repeal of the act as applied to the project. This is because nothing in the appropriations measures stated that the project was to be completed irrespective of the requirements of the Endangered Species Act. Furthermore, "Court's appraisal of wisdom or unwisdom of particular course consciously selected by

Congress is to be put aside in process of interpreting a statute. Once meaning of enactment is discerned and its constitutionality determined, judicial process comes to an end. We do not sit as a committee of review, nor is a court vested with power of veto."

Interestingly, in a dissenting opinion, Justices Lewis Powell and Harry Blackmun held that Section 7 of the Endangered Species Act cannot be reasonably applied to a project that is completed or substantially completed when its threat to an endangered species is discovered. They argue that the word *actions* in Section 7 refers to "prospective actions"—that is, actions with respect to which the agency has reasonable decision-making alternatives still available, actions not yet carried out. In an interesting footnote to this case, on 25 September 1979, President Jimmy Carter refused to veto legislation directing that the Tellico Dam be completed and the restrictions of all federal laws be waived. The history of this controversy is described in *TVA and the Tellico Dam, 1936–1979: A Bureaucratic Crisis in Post-Industrial America,* by William B. Wheeler and Michael J. Mcdonald (Knoxville: University of Tennessee Press, 1986).

Andrus, Secretary of the Interior, et al. v. Allard et al., *1979. SCT .166 (http://www.versuslaw.com); 100 S.Ct. 318*

The secretary of the interior promulgated regulations under the Eagle Protection Act and the Migratory Bird Treaty Act, which prohibit commercial transactions in parts of birds legally killed before they came under the protection of these acts. The appellees sold "preexisting" Indian artifacts containing feathers of protected birds and were prosecuted.

The appellees filed suit in district court for declaratory relief (a binding judgment of the rights and status of the litigants even though no consequential relief is awarded) and injunctive relief (a prohibitive equitable remedy that forbids a defendant to commit some act), claiming that the acts do not forbid sale of artifacts containing bird parts obtained prior to the effective dates of the acts. If the acts and regulations do apply, then they violate the Fifth Amendment of the U.S. Constitution. The district court granted the relief sought.

A three-judge court held that the acts did not apply to preexisting legally obtained bird parts or products, because if the acts did apply to pre-act bird parts, the acts might be unconstitutional. The Supreme Court held that the Eagle Protection Act allows the possession and transportation of preexisting artifacts but bans the sale of these items. The language of the act is explicit, and when Congress amended the act in 1962, it excepted only possession and transportation. The Migratory Bird Treaty Act does not provide an exception for the sale of preexisting artifacts. The acts do not require the surrender of the artifacts or place any physical invasion or restraint on them but merely prohibit their sale and thus do not effect a taking in violation of the Fifth Amendment. This is because government regulation involves adjustment of rights for the public good, and this adjustment often involves curtailment of some potential for use or economic exploitation of private property.

Richard P. Christy et al. v. Manuel Lujan Jr., Secretary of the Interior, and United States Department of the Interior, *1989. SCT. 101 (http://www.versuslaw.com); 109 S.Ct. 3176*

A sheepherder grazed his sheep on leased land near Glacier National Park, and grizzly bears, an endangered species, from the park killed a number of his sheep. Requests for help from the park rangers and attempts to frighten the bears yielded no results. When two grizzlies emerged from the forest and approached the herders' sheep, he shot and killed one of them. He was assessed a penalty for shooting the bear. He filed suit in district court seeking to enjoin enforcement of the act against herders like himself and to resist payment of the fine. He argued that his actions in defense of his livestock were protected by the Due Process Clause of the Fifth Amendment of the U.S. Constitution and that the act resulted in an uncompensated taking of his property. The district court and the circuit court (857 F.2d 1324) rejected these claims, and the Supreme Court concurred.

Justice Byron White, in a dissenting opinion, found that a person's right to protect his or her property is a liberty deeply rooted in the nation's history and tradition and is recognized by common law. Justice White suggested that the petitioner's claim for protection under the Due Process Clause raises interesting and important questions that merit plenary review. The government-authorized uncompensated taking of the petitioner's property also raises questions about Fifth Amendment rights. The government claims that it does not own the wild animals that it protects and cannot control the conduct of these animals. But the government does make it unlawful for the petitioner to "harass, harm, [or] pursue" these animals when they come to take his property. A government edict barring one from resisting the loss of his or her property may be the constitutional equivalent of an edict taking such property.

Manuel Lujan Jr., Secretary of the Interior, v. Defenders of Wildlife et al., 1992. SCT. 84 (http://www.versuslaw.com); 112 S.Ct. 2130

The Fish and Wildlife Service and the National Marine Fisheries Service on behalf of the secretary of the interior and the secretary of commerce promulgated a joint regulation extending Endangered Species Act Section 7(a)(2) coverage to actions taken in foreign nations but subsequently limited the section's geographic scope to the United States and the high seas. The Defenders of Wildlife and other environmental organizations filed suit in district court seeking declaratory and injunctive relief. The district court dismissed the suit for lack of standing, and the court of appeals reversed (851 F.2d 1035). Upon remand, the district court denied the standing motion and ordered the secretary to publish a new rule. The court of appeals affirmed (911 F.2d 117).

The Supreme Court held that respondents invoking federal jurisdiction bear the burden of proof in showing standing. They must demonstrate, by affidavit and other evidence, that they suffered an injury in fact ("a concrete and particularized, actual or imminent invasion of a legally-protected interest, . . . not a conjectural or hypothetical one"). In this case, the respondents did not demonstrate that they suffered an injury in fact. An imminent injury is not demonstrated by affidavits of members claiming they will be denied the opportunity to observe endangered animals when they revisit project sites at some indefinite future time. They must also demonstrate that there is a causal connection between the injury and the conduct complained of (that the injury has to be "fairly traceable to the challenged action of the defendant, and not . . . the result of the independent action of some third party not before the court"). They must demonstrate that it is likely, rather than merely speculative, that the injury will be redressed by a favorable decision.

When a plaintiff is the object of an action, there is usually little question that the action or inaction has caused him or her injury and that a judgment preventing or requiring the action will redress it. But in this case, the plaintiff's asserted injury arises from the government's allegedly unlawful regulation of someone else. The court has held that when a plaintiff is not the object of government action, standing is not precluded but is more difficult to establish. The respondents in this case did not show injury and failed to demonstrate redressability. The court held that under Endangered Species Act Section 1536(a)(2), the consultation initiative to determine if a proposed agency action would impact on an endangered or threatened species lies with the federal agencies and not with the secretary.

United States v. Dion, 1986. SCT. 2559 (http://www.versuslaw.com); 106 S.Ct. 2216

The respondent, a member of the Yankton Sioux Tribe, was convicted in district court of shooting four bald eagles in violation of the Endangered Species Act and of selling carcasses and parts of eagles and other birds in violation of the Eagle Protection Act and the Migratory Bird Treaty Act. The court dismissed the charge of shooting a golden eagle in violation of the Bald Eagle Protection Act. The court of appeals (762 F.2d 674) affirmed the convictions, except those dealing with shooting bald eagles in violation of the Endangered Species Act and upheld the dismissal of the other charge, finding that members of the tribe have a treaty right (11 Stat 743) to hunt eagles for noncommercial purposes on the Yankton Reservation.

The Supreme Court held that the convictions under the Endangered Species Act were valid and affirmed the dismissal of the charge under the Eagle Protection Act. The Supreme Court found that "the provisions of an act of Congress, passed in the exercise of its constitutional authority, . . . if clear and explicit, must be upheld by the courts, even in contravention of express stipulations in an earlier treaty." The Court held that the Eagle Protection Act allows the taking, possession, and transportation of eagles for religious purposes of Indian tribes, with a permit from the secretary, and the Court found that the taking of eagles for other purposes was banned. Since the respondent does not have a treaty right to hunt eagles because of the Eagle Protection Act, he cannot use treaty right as a defense to charges under the Endangered Species Act.

Bennett et al. v. Spear et al., 1997. SCT. 29 (http://www.versuslaw.com)

The Bureau of Reclamation notified the Fish and Wildlife Service that operation of the Klamath Irrigation Project, a series of lakes, rivers, dams, and irrigation canals in northern California and southern Oregon, might affect two endangered species of fish, the Lost River sucker and the shortnose sucker. The service issued a Biological Opinion (a written statement explaining how a proposed action might affect a listed species or its habitat), which provided the bureau with an outline of reasonable and prudent alternatives that would prevent the consequence. In this case, the alternatives included maintaining minimum water levels in the Clear Lake and Gerber Reservoirs. The service also issued an Incidental Take Statement specifying the terms under which the bureau could take the species.

Irrigation districts and ranch operators that received water from the project filed suit in district court claiming that the jeopardy determination and the imposition of minimum water levels violated Section 1536 of the Endangered Species Act and constituted an implicit critical habitat determination for the species. Establishing this implicit critical habitat was in violation of Endangered Species Act Section 1533(b)(2) because it did not take into account the economic impact of the designation. They also claimed that the ruling violated the "arbitrary, capricious, an abuse of discretion" rule of the Administration Procedure Act (APA). The district court dismissed the suit because the petitioners lacked standing. The petitioners asserted that they had "recreational, aesthetic, and commercial interests." The court of appeals (63 F.3d 915) concurred and held that only plaintiffs alleging an interest in the preservation of an endangered species have a cause of action.

The Supreme Court held that the statement in Section 1540(g)(1) of the Endangered Species Act that "any person may commence a civil suit" does not limit standing only to environmentalists. The Court held that the petitioner's injury could be fairly traceable to the Biological Opinion and was redressable by a favorable judicial ruling and thus met the standing requirements of Article III of the U.S. Constitution. Section 1540(g)(1)(C) of the Endangered Species Act authorizes a suit against the secretary for an alleged failure to perform any nondiscretionary act or duty under Section 1533 (which provides for listing species as threatened or endangered and allows designation of critical habitat). The Court held that the plaintiffs did not have standing via Section 1536, which requires federal agencies to avoid actions that would jeopardize listed species or affect their habitat. The Section 1536 claims are not reviewable under Section 1540(g)(1)(C), as described previously, or 1540(g)(1)(A), which authorizes injunctive action against any person who is alleged to be in violation of the Endangered Species Act or its regulations. But Section 1536 claims are reviewable under the APA, since the Biological Opinion marks the consummation of the agency's decision-making process.

U.S. Courts of Appeals

Hill et al. v. Tennessee Valley Authority, *549 F.2d 1064*

Environmental groups and others brought suit in district court under Section 11(g) of the Endangered Species Act to enjoin the Tennessee Valley Authority from completing the Tellico Dam and impounding a section of the Little Tennessee River. The U.S. district court refused a permanent injunction.

The U.S. court of appeals held that because the Tellico dam and reservoir project would jeopardize the continued existence of the snail darter, the project was in violation of the Endangered Species Act, and the court reversed the decision of the district court and ordered it to grant the injunctive relief. The court held that terminal phases of ongoing projects were not to be excluded from "actions" of departments and agencies to be scrutinized for compliance with the Endangered Species Act.

Noncompliance with the Endangered Species Act could not be condoned on the theory that congressional approval of appropriations for the dam and reservoir project signaled legislative assent with or express ratification of an agency's position that the terminal phases of an ongoing project should not be included among the "actions" that are scrutinized for compliance with the Endangered Species Act. Advisory opinions by Congress do not have the force of law. Repeal by implication is disfavored. That is, when Congress continues to provide appropriations for a project that may violate the Endangered Species Act, does this imply that Congress is indicating that this project is not subject to the provisions of the Endangered Species Act? Courts are not granted the license to rewrite statutes based on economic or other issues. This is a matter for the legislative branch. The courts are not authorized to override the secretary of the interior by changing critical habitat decisions or listing or delisting species. The courts are also ill equipped to calculate how many dollars must be invested in a project before the value of the project exceeds that of the endangered species. The responsibility of the courts is

to preserve the status quo when endangered species are threatened and allow the legislative or executive branch the opportunity to grapple with the alternatives.

The court held that conscientious enforcement of the Endangered Species Act requires that it be taken to its logical extreme.
For example, if an endangered species were discovered the day
before impoundment of a river, the Endangered Species Act requires that the impoundment be halted. The court held that impoundment of the Little Tennessee River would jeopardize the
survival of the snail darter, and thus continued work toward impoundment violates Section 1536 of the Endangered Species Act.
(See also *Hiram G. Hill et al. v. Tennessee Valley Authority* earlier in
this chapter.)

Sierra Club et al. v. Clark, *755 F.2d 608*

Approximately 1,000 eastern timber wolves (also called gray
wolves) reside in northern Minnesota, and this population has remained stable since 1976. There are no indications that the wolf
population has exceeded its carrying capacity or that population
pressures exist that cannot be relieved by a sport season. The gray
wolf population was originally listed as endangered. The Eastern
Timber Wolf Recovery Team recommended that "depredation
control" be used where wolves were killing domestic stock. The
secretary of the interior reclassified the wolf as threatened and allowed trapping of depredating wolves. When the regulations
were litigated, the district court enjoined the U.S. Fish and
Wildlife Service from trapping wolves unless such action was necessary and directed at specific wolves that were believed to have
committed significant predation upon livestock. The trapping
was to occur within a quarter mile of the place where the predation occurred. In 1982, the service proposed allowing the sport
trapping of the wolves, in which the taking could occur within a
half mile of the site of predation, the trapping would not be limited to individual predator wolves, and no express requirement
would be made that the wolves be taken in a humane manner.

Suit was brought in U.S. district court to prevent the secretary
of the interior from issuing regulations permitting sport hunting
of the eastern timber wolf under the Endangered Species Act. The
court held that the secretary may not declare a sport-hunting season for a threatened species. The court also held that public hunting of a threatened species is prohibited by the Endangered
Species Act, unless population pressures within the animal's

ecosystem cannot be otherwise resolved. The court found that the additional regulations expanding the predation control program would lead to unnecessary taking of wolves and were thus illegal.

The court of appeals concurred with the district court that the taking of a threatened species cannot occur unless population pressures within the animal's ecosystem cannot be relieved in any other manner. The court of appeals reversed the district court's decision with regard to regulations dealing with predator control and had it review the sufficiency of the statements of the secretary under the "arbitrary and capricious" rule (5 U.S.C. 706[2][A]).

In a dissenting opinion, Judge Ross argues that regulated taking of a threatened species is not limited to cases dealing with population pressures, but is limited by the necessity that regulated taking must further the effort to bring the species to a point at which Endangered Species Act measures are no longer needed. The provisions of 16 U.S.C. 1539(a)(1)(A) allow the secretary to permit "any act" to enhance the propagation or survival of an affected species.

Defenders of Wildlife v. Endangered Species Scientific Authority, 725 F.2d 726

The bobcat (*Lynx rufus*) is found throughout the United States in sufficient numbers that it is not threatened or endangered. International trade in bobcats is governed by Appendix II of the Convention on International Trade in Endangered Species of Wild Fauna and Flora (CITES), which states that certain species "although not necessarily now threatened with extinction may become so unless trade in specimens of such species is subject to strict regulation." The Office of Scientific Authority of the Fish and Wildlife Service is the U.S. agency charged with determining if trade in a species will be detrimental to its survival. The primary responsibility for the protection and management of bobcats and the regulation of their killing rests with the states.

The district court, in a previous case (*Defenders of Wildlife v. ESSA*, 659 F.2d 168), had issued an injunction barring authorization of the export of bobcats, which was affirmed by the court of appeals. The court, in a later case (659 F.2d 178), held that the guidelines developed by the working group were invalid because they failed to consider the total bobcat population in each state and the number to be killed in a particular season in each state. The Scientific Authority was required to develop additional data before authorizing the export of bobcats. On remand, the district court permanently enjoined the defendants from authorizing exports of bobcats taken or killed subsequent to 1 July 1981 until the defendants could promulgate guidelines consistent with the court of appeals decision. Congress, in the Endangered Species Amendments of 1982 (Section 8A, Paragraph c[2]), overruled this finding, and the court vacated the injunction.

Suit was brought in U.S. district court seeking a declaratory judgment regarding the Endangered Species Scientific Authority's guidelines regarding the export of bobcats and to determine the extent to which the amendments to Section 8A overruled 659 F.2d 168. The suit also sought judgment that the management regulations implementing the Convention on International Trade in Endangered Species of Wild Fauna and Flora (CITES, see Chapter 4) were invalid. Under CITES, Article IV, Section 2, exporters of animals listed in Appendix II are authorized only to the extent that the nation's Scientific Authority determines that "such export will not be detrimental to the survival of the species." The appellants argued that Congress eliminated the need for population data, but not information about projected kill levels. The district court held that Congress overruled both requirements, and the appeals court concurred. The appeals court held that Congress mandated that "the secretary shall now use the best available biological information derived from professionally accepted wildlife management practices" in making no-detriment findings.

United States of America
v. Robert Waites Guthrie, *1995. C11. 144*
(http://www.versuslaw.com)

Guthrie was charged with taking, possessing, selling, and transporting Alabama red-bellied turtles (*Pseudemys alabamensis*) and with conspiring to sell alligator snapping turtles (*Macroclemys temminicki*). The first offense was in violation of the Endangered Species Act, and the second, of the Lacey Act Amendments of 1981 (16 U.S.C. 3371). He pleaded conditionally guilty to each charge in district court. On appeal, he argued that the Lacey Act prosecution represented an unconstitutional delegation of federal legislative authority, that state regulations promulgated under Alabama state laws violate the Alabama Constitution, and that the secretary's listing of the red-bellied turtle was invalid.

The United States Court of Appeals, Eleventh Circuit, held that the prosecution under the Lacey Act was dependent on the state's nongame species regulation (Alabama Administrative Code 22–2–92). The court had held in a previous case that the Lacey Act does not unconstitutionally delegate federal legislative authority to states or their agencies. The court held in the *Guthrie* case that the state regulation is constitutional and that the 1943 Alabama Acts (Act No. 531) give the commissioner of the Alabama Department of Conservation the authority to promulgate regulations to protect wild animals, including the alligator snapping turtle. The court rejected Guthrie's argument about the invalidity of the state law and thus did not reach the issue of whether a viable state law challenge to the underlying regulation would be a defense to prosecution under the Lacey Act.

The secretary listed the Alabama red-bellied turtle as endangered (*Federal Register,* volume 52, page 22939). No one submitted public comments opposed to the listing of the turtle. The Endangered Species Act provides for a petition process for agency review (Section 1533[b][3][A]) and also allows citizen suits (Section 1540[g]). The sole defense would be that the Alabama red-bellied turtle is not a species. Guthrie never presented evidence to the secretary that the turtle is not a species, nor did he petition for further review at the time of rule making. The court held that a review is limited to the evidence before an agency at the time and that Guthrie cannot bypass the agency and receive judicial review of a regulation based on a new DNA study. The court also held that the secretary did not act in an arbitrary and capricious manner when he listed the turtle. The secretary did acknowledge that the taxonomic status of the turtle has been questioned but listed studies finding that the turtle is a valid species.

State Courts

Planning and Conservation League et al. v. Department of Fish and Game et al., 1997. CA. 285 (http://www.versuslaw.com)

The Planning and Conservation League filed suit challenging whether the California legislature, under the California Endangered Species Act (CESA Fish and Game Code, Section 2050), gave the Department of Fish and Game the authority to issue a California Endangered Species Act Permit for Emergency Measures. This permit would allow persons and public agencies to kill or capture California species that would otherwise be protected to "prevent or mitigate an emergency or natural disaster" or to "restore any property to it pre-emergency condition." This would be allowed in counties declared to be in a state of emergency by the governor or in counties or cities in which a local emergency has been declared by the local governing body or a duly designated official. Emergencies include, but are not limited to, fire, flood, earthquake, other soil or geologic movement, riot, accident, and sabotage.

The department argues that the activities authorized under such a permit and the effects of these activities on protected species are remote, speculative, and intangible and cannot constitute imminent harm. That is, the issue is not "ripe" (an administrative decision has not been formalized and its effects felt in a concrete way by the challenging parties). The court found that the "ripeness" doctrine had no application to this case.

The department used Section 2081 of the California Endangered Species Act as its authority for issuing these permits. Section 2081 states: "Through permits or memorandums of understanding, the department may authorize individuals, public agencies, universities, zoological gardens, and scientific or educational institutions, to import, export, take or possess any endangered species, threatened species, or candidate species for scientific, educational, or management purposes." The department argues that "management purposes" should be interpreted to authorize "incidental takes" of protected species in connection with any project or lawful activity that the department reviews or supervises. The Planning and Conservation League argues that the activities allowed under the permit do not "manage" or "benefit the species," but rather cause destruction to the species without any management purpose. The federal Endangered Species Act allows incidental take permits to be issued if "(1) the taking will be incidental; (2) the applicant will minimize and mitigate the impacts to the extent possible; (3) adequate funding will be insured; and (4) the taking will not appreciably reduce the likelihood of survival of the species." The department argues that the "management purposes" described previously are the functional equivalent of federal Endangered Species Act provisions. The court found that administrative bodies and their officers have only such powers as have been expressly or implicitly applied to them by the Constitution or statute, and an administrative agency may not, under the guise of regulation, substitute its judgment for that of the legislature. Without statutory authority, the administrative action must be declared void.

Mangrove Chapter of Izaak Walton League of America, Inc., and Friends of the Everglades, Inc., v. Florida Game and Fresh Water Fish Commission and Driscoll Properties, Inc., and Driscoll Foundation, Inc., 1992. FL. 115 (http://www.versuslaw.com); 592 So. 2d 1162; 17 Fla. Law W. D 228

The Florida Game and Fresh Water Fish Commission issued a permit to Driscoll Properties and the Driscoll Foundation to develop Harbor Course South. The project would allow the destruction of the nests and habitat of the Key Largo cotton mouse *(Peromyscus gossypinus allapaticola)* and the Key Largo wood rat *(Neotoma floridana smalli)*, two endangered species, incidental to land-clearing operations and building construction. The permittees would provide certain land as an alternative habitat and enhance other sites for the rodents' benefit.

The court found that the Florida Game and Fresh Water Fish Commission actions of issuing the permit with mitigation requirements set forth as conditions to the issuance of the permit are consistent with achieving the objectives of the rule.

The People ex rel. Michael Witte, Director, Illinois Department of Conservation v. Big Creek Drainage District No. 2 (David Diehl, Intervenor), 1987. IL. 1100 (http://www.versuslaw.com); 512 N.E. 2d 62; 159 Ill. App. 3d 576; 111 Ill. Dec. 158

The Illinois Department of Conservation brought suit against the Big Creek Drainage District No. 2 to prevent it from engaging in activities other than minor maintenance of the Cache River Basin in Pulaski County without notice to the department.

The Lower Cache River Natural Area, a narrow, nine-mile-long area on either side of the Cache River in Pulaski County, is commonly called the Button Land Swamp. The swamp, a natural area that displays its original, historic character, is a National Natural Heritage Landmark. It contains many endangered species of plants and animals, including large cypress and tupelo trees, the Indiana bat, and the river otter.

A low structure known as the Button Land Swamp Dam was built across the Cache River on the land of David Diehl by the Citizens to Save the Cache, a not-for-profit corporation. The dam serves to keep a base level of water in the swamp during periods of low flow in the Cache River, while during periods of high water, the water simply flows over it. If the dam were not present, a portion of the swamp could become dry during a drought. The absence of water from the swamp would have disastrous consequences for the plants and animals that live there.

The court found that the dam has little effect on flooding and that removing it could have permanent adverse consequences on the swamp and thus granted injunctive relief to the Department of Conservation.

Spotted Owl Cases

The northern spotted owl *(Strix occidentalis caurina)* is a medium-size, dark brown owl, with white spots on its head and neck, as well as mottling on its breast. It lives in old-growth and mixed old-growth/mature forests in California, Oregon, Washington, and British Columbia. The spotted owl is at the center of a divisive (and sometimes violent) controversy between environmentalists and the timber industry over the ultimate fate of the ancient forests of the Northwest. *The Northern Spotted Owl v. Hodel (716 F. Supp. 479)* was one of the first cases to deal with this controversy (see *Northern Spotted Owl v. Hodel* later in this chapter). The northern spotted owl was listed as threatened on 26 June 1990 (see *Federal Register,* volume 55, pages 26114 and 26118, Determination of Threatened Status for the Northern Spotted Owl, Final Rule) under the Endangered Species Act. The spotted owl has been the subject of dozens of lawsuits, mostly centering on resolving how, when, or if the old-growth forest that is the habitat of the owl will be harvested. Much of the controversy and the suits occur because Congress attaches case-specific riders to appropriation bills. These riders avoid committee review and public scrutiny, allowing limited harvesting of specific portions of the forest.

U.S. Supreme Court

F. Dale Robertson, Chief, United States Forest Service, et al. v. Seattle Audubon Society et al., 1992. SCT. 39 (http://www.versuslaw.com); 112 S.Ct. 1407

The Seattle Audubon Society and other environmental groups and the Washington Contract Loggers Association and other industry groups entered separate suits in district court. They claimed that a 1988 amendment to a regional Forest Service Guide that would prohibit timber harvesting in forests in Oregon and Washington that are home to the northern spotted owl, which is listed as endangered (*Federal Register,* volume 55, page 26114), provided too little and too much protection to the spotted owl, respectively. The district court consolidated the actions and enjoined the proposed timber sales.

Because of ongoing litigation, Congress enacted the Department of the Interior and Related Agencies Appropriations Act in 1990 (103 Stat 745), which is commonly called the Northwest Timber Compromise. The act expired on the last day of fiscal year 1990, but the timber sales offered under the act remained subject to its terms for the duration of the sales contracts. The compromise required both harvesting and expanded harvesting restrictions. The controversy centers on the first sentence of Subsection b(6) of this act, "The Congress hereby determines and directs that the management of areas according to subsections (b)(3) and (b)(5) of this section on the thirteen national forests in Oregon and Washington and Bureau of Land Management lands in western Oregon known to contain northern spotted owls is adequate consideration for the purpose of meeting the statutory requirements that are the basis for the consolidated cases." The district court held that Subsection b(6) "can and must be read as a temporary modification of the environmental laws" and found that the provision was constitutional. The appeals court consolidated the appeals and held (914 F.2d 1311) that Section b(6) does not repeal or amend environmental law, but rather "directs the court to reach a specific result and make certain factual findings," and thus the section is unconstitutional.

The Supreme Court held that subsection (b)(6)(A) compelled changes in the law. It does not compel findings or results under old law, but rather modifies the old provisions. This subsection does not direct any particular findings or fact or applications of old law in existing cases before the courts.

Babbitt, Secretary of Interior, et al. v. Sweet Home Chapter of Communities for a Great Oregon et al., 1995. SCT. 141 (http://www.versuslaw.com)

Small landowners, logging companies, and families dependent on forest products industries, as well as organizations that represent their interests, challenged the statutory validity of a regulation promulgated by the secretary of the interior (50 CFR 17.3) that defines *harm* to include "significant habitat modification or degradation where it actually kills or injures wildlife by significantly impairing essential behavioral patterns" and sought declaratory judgment relief in district court. The district court considered each of the respondent's arguments and rejected them (806 F. Supp. 279), a decision that was reversed by the court of appeals (17 F.3d 1463). The court of appeals held that the immediate statutory context in which *harm* appeared argued against a broad reading and, like other words in the definition of *take*, should ally only to the "perpetrator's direct application of force against the animal taken." This decision created a square conflict with *Palila v. Hawaii Department of Land and Natural Resources* (852 F.2d 1106).

The respondents provided three arguments that Congress did not intend the word *take* in Section 9 of the Endangered Species Act to include habitat modification. First, the Senate eliminated language that defined *take* to include habitat modification from its original version of the bill. Second, Congress intended that the authorization to buy private land was the exclusive check to prevent habitat modification on private property. Third, the Senate added the term *harm* to the definition of *take* in a floor amendment without debate.

The Supreme Court in making its judgment assumed that the respondents had no desire to harm the red-cockaded woodpecker or the northern spotted owl, the two endangered species involved, but merely wanted to continue logging operations, if they were not prohibited by the Endangered Species Act. These activities will, although unintentionally, alter the listed species detrimentally and cause them to be harmed or killed. The Court held that there are three reasons for accepting the secretary's interpretation. First, the ordinary definition of *harm*—that is, to cause hurt or damage—would encompass habitat modification, as it would result in actual injury or death to members of a listed species. If the word *harm* means only direct applications of force against a listed species, then it does not have any meaning that does not duplicate the meaning of other words in Section 3 of the Endangered Species Act. Second, the broad purpose of the Endangered Species Act is to protect ecosystems upon which listed species depend. Third, Congress authorized the secretary to issue permits that would allow incidental taking, and thus Congress understood that Section 9(a)(1)(B) of the Endangered Species Act prohibits indirect as well as direct takings.

The Court held that several words that accompany *harm*, such as *harass* and *pursue*, refer to actions that do not require the direct application of force. Further, a "knowing" action is enough to violate the Endangered Species Act. The Court held that the appeals court was wrong to use *noscitur a sociis* (that is, "it is known from its associates," meaning that the meaning of a word is or may be known from the accompanying words) to give *harm* the same function as other words in the definition of *take*, since that would deny the word *harm* its own independent meaning. *Noscitur a sociis* counsels that a word "gathers meaning from the words around it," and the Court held that Congress meant *harm* to have a function distinct from that of the other verbs in the meaning of *take*. The secretary's interpretation of *harm* to include indirectly injuring listed species through habitat modification is appropriate. The provision for buying private land does not alter this conclusion. The legislative history of the act supports the secretary's interpretation.

U.S. Courts of Appeals

Marbled Murrelet (*Brachyramphus marmoratus*); Northern Spotted Owl (*Strix occidentalis caurina*); Environmental Protection Information Center v. Bruce Babbitt, Secretary, U.S. Department of Interior, and Pacific Lumber Company; Scotia Pacific Holding Company; Salmon Creek Corporation, 1997. C09. 1051 (http://www.versuslaw.com)

The Environmental Protection Information Center (EPIC) filed suit in district court on behalf of the marbled murrelet and the northern spotted owl, seeking an injunction enjoining the appellants from conducting logging activity in Humboldt County, California. The district court provided injunctive relief based on its view that EPIC demonstrated that the Endangered Species Act and the National Environmental Protection Act may have been violated. The appellants argued that EPIC failed to give 60 days notice of intent to sue under Section 7 of the Endangered Species Act as required by Section 11(g) of the Endangered Species Act. EPIC responded that the intent to sue was given in letters written to the appellants, the U.S. Fish and Wildlife Service, and the California Department of Forestry and Fire Protection. In California, Timber Harvest Plans must be approved by the California Department of Forestry, and the plan submitters must provide information to the director of the department, which he or she uses to determine if the proposed activity will result in the "take" of an individual spotted owl. The plan submitters provided the Fish and Wildlife Service with concurrence letters opining that the planned timber operations would not likely result in the "take" of the owl. The appeals court held that EPIC did not raise serious questions as to whether the Fish and Wildlife Service engaged in agency action when it issued the concurrence letters.

Maricopa County Audubon Society and Robin Silver v. United States Forest Service, 1997. C10. 17 (http://www.versuslaw.com)

The Maricopa County Audubon Society and Robin Silver filed suits in district court to obtain maps identifying specific Mexican spotted owl nest sites in national forests in New Mexico and Arizona. The Mexican spotted owl is an Endangered Species Act threatened species. The plaintiffs had requested the maps from the Forest Service under the Freedom of Information Act (FOIA). The Forest Service denied the request, claiming that the maps could be withheld under Exemption 2 of FOIA (5 U.S.C. 552[b][2]), which allows an agency to withhold information that is "related solely to the internal personnel rules and practices of the agency." In the district court case, the Forest Service, under the "high 2" interpretation of the exemption (see, for example, 964 F.2d 1205), claimed that the maps fall within the statutory language. The release of the maps would make it easier to find and harm the owls. The district court did not accept the "high 2" interpretation and required the plaintiffs to enter into a confidentiality agreement with the Forest Service not to reveal the maps to anyone not named in the agreement.

The federal agency resisting disclosure bears the burden of justifying nondisclosure. The "high 2" approach allows exemption if (1) the information falls within the language of the exemption and (2) its disclosure would risk circumvention of federal statutes or regulations. The appeals court held that the phrases "internal personnel rules" and "practices of an agency" should not be read disjunctively, that "internal personnel" modified both "rules" and "practices." Therefore, the question is not whether the owl maps relate to "agency practices," but rather whether they relate to the "personnel practices" of the Forest Service. The court held that it stretches the language of the exemption too far to conclude that the maps relate to the personnel practices of the Forest Service. Therefore, the argument fails the first prong of the "high 2" test, and the court did not consider the second argument. The court also did not address the confidentiality issue, because the plaintiffs did not challenge it.

Seattle Audubon Society v. Robertson, *914 F. 2d 1311*

The Seattle Audubon Society and other environmental groups sued in district court seeking declaratory and injunctive relief and challenging that the forest management practices of the Bureau of Land Management violated the National Environmental Protection Act (NEPA) (42 U.S.C. 4321–4347), the Federal Land Policy and Management Act (43 U.S.C. 1701–1782), the Migratory Bird Treaty Act (16 U.S.C. 703–711), as well as the Oregon and California Lands Act (43 U.S.C. 1181). The district court dismissed the action and was reversed by the court of appeals (866 F.2d 302) and remanded to the district court. The district court dismissed the action, and the court of appeals affirmed the NEPA claims but reversed the non-NEPA claims (884 F.2d 1233). The Seattle Audubon Society renewed its non-NEPA claims in district court, and the Seattle Audubon Society filed suit in district court for declaratory and injunctive relief because the Forest Service timber management guidelines did not afford adequate protection for the spotted owl. The Washington Contract Logging Association filed suit in district court challenging the guidelines as overly restrictive of timber harvesting.

Congress enacted Section 318 of the Department of Interior and Related Agencies Appropriations Act for Fiscal Year 1990 (Public Law 101–121, 103 Stat 701, 745–750), commonly called the Northwest Timber Compromise, which would require the sale of 5.8 billion board feet of timber from the old-growth forests of Oregon and Washington. None of the proposed timber sales were to come from spotted owl habitat areas identified in the environmental impact statement and the Record of Decision of 1988. Congress "hereby determines and directs that management of area according to subsections (b)(3) and (b)(5) of this section [that is, 318] on the thirteen national forests in Oregon and Washington and Bureau of Land Management lands in western Oregon known to contain northern spotted owls is adequate consideration for the purpose of meeting the statutory requirements that are the basis for the consolidated cases."

Based on Section 318 of the Department of the Interior and Related Agencies Appropriations Act of 1990, the district court vacated the preliminary injunction granted under Seattle and rejected the arguments that Section 318 violated the powers doctrine and was unconstitutional. The district court retained jurisdiction. The district court granted the government's motion in Portland of constitutionality.

The question before the appellate court was whether Congress, in enacting Section 318, went beyond its constitutional powers when it instructed federal courts to reach a particular result in cases identified by caption and file number.

The appeals court held that Section 318 does not, by its plain language, repeal or amend the environmental laws underlying this litigation (which it can do), but prescribes a rule for the decision of a case in a particular way, without changing the underlying law (which it cannot do). The appeals court reversed the district court's decision and remanded the cases (the Seattle Audubon Society case and the Washington Contract Logging Association case) to the district courts.

U.S. District Court

Northern Spotted Owl v. Hodel, 716 F. Supp. 479

A number of environmental groups brought suit in district court against the U.S. Fish and Wildlife Service (FWS) claiming that the decision not to list the northern spotted owl as endangered or threatened under the Endangered Species Act was arbitrary and capricious or contrary to law.

In January 1987, Greenworld petitioned the FWS to list the spotted owl as endangered, and in August 1987, 29 conservation groups filed a second petition to list the owl as endangered in the Olympic Peninsula in Washington and in the Oregon Coast Range and as threatened throughout the rest of its range. In July 1987, the FWS announced that it would initiate a status review of the spotted owl. After conducting this review, the FWS concluded that listing the owl as endangered was not warranted (*Federal Register,* volume 52, pages 48552 and 48554). That decision led to this suit, with both sides requesting summary judgment. In cases like this, the judge is required to view the evidence in the light most favorable to the nonmoving party. Under the arbitrary and capricious claim, the court must determine if the agency decision was "based on consideration of the relevant factors" and whether it engaged in a "substantial inquiry." The FWS Status Review and Finding did not offer much insight into how the FWS found the owl currently has a viable population. The Status Review cites extensive empirical data and lists conclusions but does not provide any analysis. The court held that the FWS did not set forth grounds for its decision not to list the owl and also held that the expert analysis is entirely to the contrary. In deference to FWS expertise, the court remanded the matter to the FWS and allowed 90 days for the FWS to provide an analysis of its decision.

Organizations 6

Numerous organizations are involved in activities that are affected by the controversy surrounding endangered species. The controversy is a major reason for the existence of some groups, whereas it is only a minor portion of the focus of others. This chapter, while not exhaustive, provides an annotated list of organizations affected to a greater or lesser degree by the issue of endangered species.

The first section of this chapter lists groups that focus broadly on aspects of conservation and biodiversity. The second section lists organizations that focus on specific species or types of organisms.

General Organizations

Abundant Wildlife Society
of North America (AWS)
12665 Highway 59 N
Gillette, WY 82716
Phone: (307) 682–2826

The AWS supports multiple uses of public lands, including grazing, logging, mining, hunting, fishing, trapping, and recreation. The organization promotes management and use of wildlife, including predator control, but does not endorse the needless destruction of habitat. The AWS keeps members apprised of environmental and animal rights agendas.

Publications: Abundant Wildlife, bimonthly newsletter. Brochures. Reports. Books: *Seven Popular Myths about Livestock Grazing on Public Lands; Trashing the Planet; Unnatural Wolf Transplant Yellowstone National Park; Wolf Reintroduction in Yellowstone National Park: A Historical Perspective.*

African Wildlife Foundation (AWF)
1717 Massachusetts Avenue NW
Washington, DC 20036
Phone: (202) 265–8393
Fax: (202) 265–2361

Founded in 1961, the AWF trains Africans in wildlife management and ecology in its schools of wildlife management in Tanzania and Cameroon. The organization offers technical assistance in park and reserve management and provides vehicles, aircraft, tents, uniforms, radios, and training to antipoaching patrols. The AWF supports management research, works in conjunction with other conservation organizations, and initiates and supports conservation education throughout Africa.

Publications: Wildlife News, quarterly.

Alliance for Environmental Education Inc. (AEEI)
P.O. Box 368
The Plains, VA 22171
Phone: (703) 253–5812

The AEEI was organized in 1972 and is made up of many diverse regional, national, and international groups, including conservation and education organizations, corporations, and organized labor. The AEEI coalition advocates environmental education in North America and around the world.

Publications: Network Exchange.

**American Association of Botanical Gardens
and Arboreta (AABGA)**

Organized in 1940, AABGA is a nonprofit membership organization serving North American botanical gardens and arboreta and their professional staffs.

Publications: The Public Garden; AABGA Newsletter.

American Association of Zoo Keepers (AAZK)
Administrative Offices
635 S.W. Gage Boulevard
Topeka, KS 66606–2066
Phone: (913) 272–5821, ext. 31
Fax: (213) 273–1980

Established in 1967, the AAZK is an international nonprofit organization of animal keepers and other people interested in high-quality animal care and in promoting animal keeping as a profession. The AAZK provides continuing education for keepers, promotes national and international conservation projects, supports keeper-initiated zoo research, and puts out educational publications. The group has chapters at zoos throughout North America.

Publications: Animal/Keepers' Forum magazine; Biological Values for Selected Mammals, Diet Notebook/Mammals vol. 1; Handbook of Zoonotic Diseases.

American Association of Zoological Parks and Aquariums (AAZPA)
Executive Office/Conservation Center
7970-D Old Georgetown Road
Bethesda, MD 20814
Phone: (301) 907–7777
Fax: (301) 907–2980

Organized in 1924, AAZPA is dedicated to the improvement of zoological parks and aquariums. AAZPA advocates professional management, conservation, public education, and scientific research, and the group administers scientifically managed captive breeding and field conservation programs for 69 threatened and endangered species. AAZPA is a member of the International Union for the Protection of Nature and Natural Resources—the World Conservation Union.

Publications: Communique; Directory of Zoological Parks and Aquariums in the Americas.

American Committee for International Conservation (ACIC)
Center for Marine Conservation
1725 DeSales Street NW, Suite 500
Washington, DC 20036
Phone: (202) 429–5609
Fax: (202) 872–0619

Founded in 1930, ACIC strives to promote conservation and the preservation of wildlife and other natural resources. ACIC encourages and finances research on the status and ecology of threatened species and maintains a working relationship with the International Union for Conservation of Nature and Natural Resources.

American Conservation Association (ACA)
30 Rockefeller Plaza, Room 5402
New York, NY 10112
Phone: (212) 649–5600
Fax: (212) 649–5921

Founded in 1958, the ACA acts as a clearinghouse for information on the environment, in an effort to raise public awareness of environmental and conservation issues. The group sponsors programs to preserve and develop natural resources.

American Ecological Research Institute (AERIE)
P.O. Box 380
Fort Collins, CO 80522
Phone: (303) 224–5307

AERIE was organized in 1985 as an international natural resource consultation and research firm. The organization does research, education, consultation, training, and conservation, with emphasis on nonintrusive field studies and an interdisciplinary ecological approach. AERIE focuses on threatened and endangered species, wildlife disease and toxicology, the study of tracks and signs, and cryptozoology.

Publications: Eastern Panther Update.

American Fisheries Society (AFS)
5410 Grosvenor Lane, Suite 110
Bethesda, MD 20814
Phone: (301) 897–8616

Organized in 1870, the AFS is a professional society set up to promote the conservation, development, and wise utilization of fisheries, both recreational and commercial.

Publications: Transactions of the American Fisheries Society; North American Journal of Fisheries Management; Progressive Fish-Culturist; Journal of Aquatic Animal Health.

American Forest and Paper Association (AFC)
1250 Connecticut Avenue NW, Second Floor
Washington, DC 20036
Phone: (202) 463–2455
Fax: (202) 463–2461

The AFC was founded in 1932 as the national trade association of the forest, pulp, paper, paperboard, and wood products industry.

Publications: American Tree Farmer: The Official Magazine of the American Tree Farm System, bimonthly; *The Branch: The Official Magazine of Project Learning Tree; In Focus,* monthly; *International Trade Report; This Month,* monthly newsletter; *This Week,* weekly newsletter; *Paper, Paperboard and Wood Pulp Capacity and Fiber Consumption,* annual; *Statistical Summary of Recovered Paper Utilization,* annual; *Statistics of Paper, Paperboard and Wood Pulp,* annual.

American Forest Council (AFC)
1250 Connecticut Avenue NW, Suite 320
Washington, DC 20036
Phone: (202) 463–2455

Organized in 1932, the AFC, a trade organization, strives to ensure long-term growth, quality, and availability of private forest resources and to foster public understanding and acceptance of productive forest stewardship.

Publications: American Tree Farmer magazine; *PLT Branch, Forestry Communications Catalogue.*

American Forest Foundation (AFF)
1250 Connecticut Avenue NW, Suite 320
Washington, DC 20036
Phone: (202) 463–2455

The AFF was organized in 1981 to support the American Tree Farm System and Project Learning Tree, a K–12 environmental education curriculum and training program. The American Forest Council administers AFF programs.

Publications: The PLT Activity Guide (K–6 and 7–12); GreenAmerica poster series.

American Forests (AF)
1516 P Street NW
Washington, DC 20005
Phone: (202) 667–3300

AF, founded in 1875, is a citizens' conservation organization that strives to promote public appreciation of natural resources and the part they play in the social, recreational, and economic life of the United States. AF works to advance the intelligent management and use of forests, soil, water, wildlife, and all other natural resources. Until 1992, the group was known as the American Forestry Association.

Publications: American Forests, Urban Forests, bimonthly magazines. Books: *National Registry of Champion Big Trees and Famous Historical Trees.* Reprints. *Resource Hotline,* biweekly newsletter.

American Friends of the Game Conservancy (AFGC)
910 Pierremont Road, Suite 250
Shreveport, LA 71106
Phone: (318) 868–3631

The AFGC, organized in 1985, raises funds to support the research and educational activities of the Game Conservancy Trust in the United Kingdom.

Publications: American Friends of the Game Conservancy.

American Friends of the Wildfowl
and Wetlands Trust (AFWWT)
20110 U.S. 12
White Pigeon, MI 29099–9750
Phone: (616) 651–6417
Fax: (616) 651–3679

Founded in 1990, the AFWWT works to protect wildfowl and wetlands.

Publications: American Friends, quarterly newsletter.

American Littoral Society (ALS)
Sandy Hook
Highlands, NJ 07732
Phone: (201) 291–0055

The ALS, organized in 1961, conducts field trips, dive and study expeditions, and a fish tag-and-release program, as well as activities for scuba divers and underwater photographers. The organization supports the study and conservation of coastal habitat, barrier beaches, wetlands, estuaries, and near-shore waters and those areas' fish, shellfish, bird, and mammal resources.

Publications: Underwater Naturalist; Coastal Reporter.

American Society of Zoologists (ASZ)
104 Sirius Circle
Thousand Oaks, CA 91360
Phone: (805) 492–3585

The ASZ, organized in 1890, is an association of professional zo-ologists that supports public dissemination of new and impor-tant facts and concepts in the area of animal biology.

Publications: The American Zoologist.

American Wildlands (AWL)
7500 E. Arapahoe Road, Suite 305
Englewood, CO 80111
Phone: (303) 773–1804

Founded in 1977, AWL seeks to promote the protection and re-sponsible management of wildland resources, including wilder-ness, watersheds, wetlands, free-flowing rivers, fisheries, forests, and wildlife. The group seeks to identify and investigate wilder-ness areas, wild and scenic rivers, and other natural areas need-ing protection. AWL sponsors forums, institutes, and other programs on proper land and water management on publicly owned lands, including the Timber Management Policy Reform and Sustainable Forestry Program, which seeks reform on na-tional forest policy. The organization assists citizens in preparing proposals to add areas to the national wilderness system, the na-tional wild and scenic river system, and other land and water special management systems. Until 1989, the AWL was known as the American Wilderness Alliance.

Publications: The Citizen Forester, newsletter; *The Northern Rockies Forest; On the Wild Side,* quarterly journal.

Americans for the Environment (AFE)
1400 16th Street NW, Box 24
Washington, DC 20036–2266
Phone: (202) 797–6665
Fax: (202) 797–6645
eng.org/pub/home/afe/homepage.htm

The AFE was organized in 1982 to provide nonpartisan educa-tion to help citizen activists use the political process to help solve environmental problems. The AFE does not support individual candidates.

Publications: The Rising Tide: Public Opinion, Policy and Politics; Taking the Initiative; The Power of the Green Vote; The Political Agenda of the "Wise Use" Movement.

AMNET
16056 E. Columbia Place
Aurora, CO 80013
Phone: (303) 680–9011
Fax: (303) 680–7791

AMNET was founded in 1983 to promote positive ethical concepts of the human relation to animals and the environment.

Publications: Brochure.

Ancient Forest International (AFI)
P.O. Box 1850
Redway, CA 95560
Phone: (707) 923–3015
Fax: (707) 923–3015

AFI, founded in 1989, strives to document the distribution of ancient rain forests worldwide and promote their preservation, as well as increase awareness of Earth's few still-intact temperate forest ecosystems.

Publications: News of Old Growth; Chile's Native Forest: An Overview.

Association for Conservation Information (ACI)
P.O. Box 12559
Charleston, SC 29412
Phone: (803) 762–5032
Fax: (803) 762–5007

The ACI, founded in 1938, is a professional society of information and education personnel of state, provincial, federal, and private conservation agencies.

Publications: The Balance Wheel, quarterly; membership directory, annual.

**Association of Environmental
and Resource Economists (AERE)**
1616 P Street NW, Room 507
Washington, DC 20036

Phone: (202) 328–5077
Fax: (202) 939–3460

Founded in 1981, AERE is concerned with issues such as water and land resources and air pollution, as well as problems and concerns in resource management.

Publications: AERE Newsletter, semiannual; *Journal of Environmental Economics and Management,* bimonthly.

Boone and Crockett Club (B&C)
Old Milwaukee Depot
250 Station Drive
Missoula, MT 59801
Phone: (406) 542–1888
Fax: (406) 542–0784

B&C, founded in 1887, strives to preserve North American big game for hunters. The group maintains a library and an archive. It also sponsors educational programs and graduate-level research and workshops on wildlife species.

Publications: An American Crusade for Wildlife; The Black Bear in Modern North America; Boone and Crockett Club's 21st Big Game Awards; Measuring and Scoring North American Big Game Trophies; Records of North American Big Game; Records of North American Elk and Mule Deer; Records of North American Whitetail Deer; The Wild Sheep in Modern North America; Boone and Crockett Club News Journal, quarterly.

The Camp Fire Club of America (CFCA)
230 Camp Fire Road
Chappaqua, NY 10514
Phone: (914) 941–0199
Fax: (914) 923–0977

The CFCA, organized in 1897, strives to preserve forests and woodland, as well as to protect and conserve the wildlife of our country.

Publications: The Backlog.

Carrying Capacity Network (CCN)
1325 G Street NW, Suite 1003
Washington, DC 20005–3104
Phone: (800) 466–4866, (202) 879–3044

Founded in 1989, the CCN focuses on sustainable development issues, including environmental protection, population stabilization, growth control, and resource conservation.

Publications: Clearinghouse Bulletin; Focus.

Center for Environmental Information (CEI)
50 W. Main Street
Rochester, NY 14614
Phone: (716) 262–2870

The CEI, founded in 1974, provides on-call reference and referral, current information, and educational services to scientists, educators, government agency staff, policy makers, business and industry managers, and interested citizens. The organization maintains its own library collection.

Publications: Global Climate Change Digest; proceedings of annual conferences.

Center for Environmental Philosophy (CEP)
Chestnut Hall
1926 Chestnut Street, Suite 14
P.O. Box 13496
Denton, TX 76203–6496
Phone: (817) 565–2727
Fax: (817) 565–4448

The CEP, founded in 1980, promotes research and instruction in environmental ethics and its application in environmental policy and decision making.

Publications: Environmental Ethics.

Center for Environmental Study (CES)
Grand Rapids Community College
143 Bostwick NE
Grand Rapids, MI 49503
Phone: (616) 771–3935
Fax: (616) 771–4005

Founded in 1968, the CES provides research, education, information/communication, and consulting services related to global, national, and regional environmental issues and programs.

Center for International Environmental Law (CIEL)
1621 Connecticut Avenue NW, Suite 200

Washington, DC 20009–1076
Phone: (202) 332–4840

The CIEL, founded in 1989, is a public-interest law firm specializing in international environmental law.

Publications: State of Environmental Law; CIEL Newsletter.

Center for Marine Conservation (CMC)
1725 DeSales Street NW, Suite 500
Washington, DC 20036
Phone: (202) 429–5609
Fax: (202) 872–0619

Founded in 1972, the CMC has programs in fisheries conservation, species recovery, pollution prevention, habitat conservation, and marine biological diversity. Programs focus on research, policy analysis, education, and public information and involvement.

Publications: Marine Conservation News; Sanctuary Currents; Coastal Connection. List of additional publications available on request.

Center for Plant Conservation (CPC)
P.O. Box 299, St. Louis, MO 63166
Phone: (314) 577–9450
Fax: (314) 577–9465
mobot.org/cpc/

The CPC is a national network of botanical gardens and arboreta dedicated to the conservation and study of rare and endangered U.S. plants.

Publications: Plant Conservation, newsletter; *Plant Conservation Directory.*

Charles Darwin Foundation for the Galapagos Isles (CDFGI)
National Zoological Park
Washington, DC 20008
Phone: (202) 673–4705
Fax: (703) 538–6835

The CDFGI, founded in 1959, supports, organizes, and administers research work on science and conservation at a station authorized by the government of Ecuador for the protection of the wildlife of the Galápagos Islands. The organization maintains a museum of Galápagos biological specimens.

Publications: Annual report (in English and Spanish); *Noticias de Galápagos*, semiannual.

Conservation and Research Foundation (CRF)
P.O. Box 3420
Kansas City, KS 66103–0420
Phone: (913) 268–0076

Founded in 1953, the CRF seeks to encourage biological research, promote conservation of renewable natural resources, and deepen the understanding of the relationship between humans and the environment.

Publications: Five-year report.

The Conservation Fund (CF)
1800 N. Kent Street, Suite 1120
Arlington, VA 22209
Phone: (703) 525–6300

The CF, founded in 1985, is dedicated to advancing land and water conservation with creative ideas and new resources.

Publications: Land Letter; Common Ground.

Conservation International (CI)
1015 18th Street NW, Suite 1000
Washington, DC 20036
Phone: (202) 429–5660
Fax: (202) 887–5188

Founded in 1987, CI works with governments and other organizations to help all nations develop the ability to sustain biological diversity and the ecosystems that support life on Earth while addressing basic economic and social needs.

Publications: The Debt for Nature Exchange; Orion Nature Quarterly. Published in conjunction with Myrin Institute. *The Rain Forest Imperative: A Ten-Year Strategy to Save Earth's Most Threatened Ecosystems; Tropicus,* quarterly newsletter.

Conservation Treaty Support Fund (CTSF)
3705 Cardiff Road
Chevy Chase, MD 20815
Phone: (301) 654–3150
Fax: (301) 652–6390

The CTSF, founded in 1986, seeks to promote public awareness and understanding of conservation treaties and their goals, as well as to enhance public support for, compliance with, and funding of these treaties. The organization maintains the Conservation Treaty Support Force and Convention on International Trade in Endangered Species of Wild Fauna and Flora Ambassadors Club.

Publications: CITES Junior Patrol.

The Cousteau Society (CS)
870 Greenbrier Circle, Suite 402
Chesapeake, VA 23320
Phone: (804) 423–9335

Founded in 1973, the CS strives to encourage environmental protection and highlight the need to ensure ecological sustainability for present and future generations. The group believes that an informed and alerted public can best make the choices that will promote global stability. The CS produces television films, research, lectures, books, and other publications.

Publications: Calypso Log; Dolphin Log.

Defenders of Wildlife (DW)
1244 19th Street NW
Washington, DC 20036
(202) 659–9510

DW, founded in 1947, believes that wildlife has an intrinsic value and that wildlife offers many benefits to society. The organization advocates governmental, citizen, and legal action on behalf of endangered species, habitat conservation, predator protection, and wildlife appreciation.

Publications: Defenders.

The Desert Protective Council (DPC)
P.O. Box 4294
Palm Springs, CA 92263

Founded in 1954, the DPC seeks to safeguard by "wise and reverent use" those desert areas that are of unique scenic, scientific, historical, spiritual, and recreational value.

Publications: El Paisano, newsletter.

Earth Day 2000 (ED)
116 New Montgomery Street, Number 530
San Francisco, CA 94105
Phone: (800) 727–8619, (415) 495–5987
Fax: (415) 543–1480

ED, founded in 1991, provides reports on products and services
that are good for the environment. The group also campaigns for
truth in environmental claims on labeling and in advertising.

Publications: Countdown 2000, annual report; *Earth Day 2000,* bimonthly newsletter.

Earth Day U.S.A. (EDUSA)
2 Elm Street, Box 470
Peterborough, NH 03458
Phone: (603) 924–7720
Fax: (603) 924–7855

EDUSA coordinates annual Earth Day events and activities in the
United States.

Earth Ecology Foundation (EEF)
612 N. Second Street
Fresno, CA 93702
Phone: (209) 442–3034

The EEF, founded in 1980, strives to study the interrelationships
between humans, technology, and nature; to provide modern solutions to ecological problems; and to develop disaster management plans pertaining to climate change. The organization
promotes the scientific and humane use of Earth's ecology
through both natural and technological means.

Publications: Earth Echo, quarterly newsletter; *Earthology—The
Solar Pattern Governing Life on Earth.*

Earth First! (EF!)
P.O. Box 5176
Missoula, MT 59806

EF! was founded in 1980, as a radical ecology group, to promote
a biocentric worldview and the preservation of natural diversity.
The group uses the slogan "No compromise in the defense of
Mother Earth!"

Publications: Earth First! Journal in Defense of Wilderness and Biodiversity, eight issues per year. Also publishes summaries of research and distributes calendars, T-shirts, hats, books, and bumper stickers.

Earth Island Institute (EII)
300 Broadway, Suite 28
San Francisco, CA 94133
Phone: (415) 788–3666

The EII, founded in 1982, strives to develop innovative projects that promote the conservation, preservation, and restoration of the global environment. Projects include the International Marine Mammal Project, the Sea Turtle Restoration Project, the Urban Habitat Program, and Baikal Watch.

Publications: Earth Island Journal; Race, Poverty and the Environment; Dolphin Update; Sea Turtle Restoration Project Update.

EarthAction International (EI)
30 Cottage Street
Amherst, MA 01002
Phone: (413) 549–8118
Fax: (413) 549–0544

EI was founded in 1988 to mobilize international public pressure on key decision makers when important global decisions are being made.

Publications: EarthAction Alert, monthly; *EarthAction Media Alert; EarthAction Report.*

Earthwatch
P.O. Box 403N
Mount Auburn Street
Watertown, MA 02272
Phone: (800) 776–0188, (617) 926–8200

Founded in 1971, Earthwatch sponsors scientific field research worldwide. Volunteers pay to work for two weeks alongside field scientists in 50 countries, studying endangered species, rain forest ecology, animal behavior, and ozone depletion.

Publications: Earthwatch magazine.

Ecoforestry Institute—United States (EI)
P.O. Box 12543
Portland, OR 97212
Phone: (503) 231–0576
Fax: (503) 231–0576

The EI, founded in 1991, promotes environmental, productive, recreational, aesthetic, and spiritual uses of natural forests, creates demonstration forests to provide the public with examples of ecological forestry, and promotes the philosophy and practice of natural-selection ecoforestry, in which natural selection indicators are used to determine which trees to remove from the forest. The group encourages restoration of plantation tree farms.

Publications: Ecoforesters, quarterly.

The Ecological Society of America (ESA)

Founded in 1915, the ESA is North America's professional society of ecologists. The organization encourages the scientific study of organisms in relation to their environment and promotes the exchange of ideas among those interested in ecology.

Publications: Ecology; Ecological Applications; Ecological Monographs; Bulletin of the Ecological Society of America.

The Ecotourism Society (ES)
P.O. Box 755
North Bennington, VT 05257
Phone: (802) 447–2121

The ES, organized in 1990, seeks to find the resources and build the expertise to make tourism a viable tool for conservation and sustainable development.

Publications: Ecotourism: A Guide for Planners and Managers; Ecotourism Bibliography; quarterly newsletter.

Eleventh Commandment Fellowship (ECF)
1017 N. Wahsatch Avenue
Colorado Springs, CO 80903
Phone: (719) 635–6177

The ECF, founded in 1979, seeks to have members of all faiths accept and practice the Eleventh Commandment as articulated by Roderick Nash and expanded by Vincent Rossi: "The earth is the

Lord's and the fullness thereof; thou shall not despoil the earth nor destroy the life thereon." The group sponsors seminars on the philosophy and practice of Earth stewardship; conducts local ecology action programs, wilderness retreats, youth programs, and outings; and maintains a 2,000-volume library on Christian ecology.

Publications: Chapter News, periodic; *The Eleventh Commandment Newsletter,* semiannual; *The Eleventh Commandment: Toward an Ethic of Ecology.*

Endangered Species Coalition (ESC)
56 Pennsylvania Avenue SE
Washington, DC 20003
Phone: (202) 547–9009
Fax: (202) 547–9022

Founded in 1981, the ESC strives to extend the Endangered Species Act and ensure its effective implementation. The organization seeks to develop an efficient means of listing endangered or threatened species based upon the best available scientific and commercial data. The ESC also encourages international cooperation in the conservation of endangered and threatened species via the implementation of conservation agreements. The group seeks to secure a strong legal base for the effective conservation and recovery of plants and animals in danger of extinction and encourages public participation through citizens' lawsuits and petitions for the listing of particular species.

Publications: The Endangered Species Act: A Commitment Worth Keeping.

Environmental Action (EA)
6930 Carroll Avenue, Suite 600
Takoma Park, MD 20912
Phone: (301) 891–1100
Fax: (301) 391–2218

EA was founded in 1970 by the organizers of Earth Day 1970. EA's political action committee, Enact/PAC, works to elect environmentalists to Congress by making endorsements and publicizing the voting records of antienvironmentalist members of Congress.

Publications: Environmental Action.

Environmental Action Coalition (EAC)
625 Broadway, Second Floor
New York, NY 10012
Phone: (212) 677–1601

The EAC, founded in 1970, serves as a clearinghouse for environmental services in the New York City area and other urban areas nationwide. The group seeks to educate the public about the nature and scope of major environmental problems, to provide a resource center to help concerned citizens develop positive solutions to these problems, and to motivate the public to become involved in these solutions.

Publications: Cycle, quarterly newsletter; *Eco-News: An Environmental Newsletter for Children,* periodic; *Waste Paper.*

Environmental Action Foundation (EAF)
6930 Carroll Avenue, Suite 600
Takoma Park, MD 20912
Phone: (301) 891–1100
Fax: (301) 891–2218

The EAF, founded in 1970, seeks to develop research and conduct broad educational programs on complex environmental issues.

Publications: Wastelines.

Environmental and Energy Study Institute (EESI)
122 C Street NW, Suite 700
Washington, DC 20001
Phone: (202) 628–1400
Fax: (202) 628–1825

Founded in 1985, the EESI seeks to educate policy makers and the general public on issues such as groundwater protection, water efficiency, and global climate change.

Publications: Briefing Book, annual.

Environmental Defense Fund (EDF)
257 Park Avenue South
New York, NY 10010
Phone: (212) 505–2100

EDF, founded in 1967, seeks solutions to global issues such as ocean pollution, rain forest destruction, and global warming. The group uses multidisciplinary teams of scientists, economists, and attorneys to develop economically viable solutions to environmental problems.

Publications: EDF Letter.

Environmental Law Alliance Worldwide (E-LAW)
1877 Garden Avenue
Eugene, OR 97403
Phone: (503) 687–8454
Fax: (503) 687–0535

E-LAW, organized in 1989, is an international network of public-interest attorneys and scientists dedicated to using law to protect the environment.

Publications: E-LAW Update.

The Environmental Law Institute (ELI)
1616 P Street NW, Suite 200
Washington, DC 20036
Phone: (202) 328–5150, (202) 328–5002

ELI, founded in 1969, seeks to advance environmental protection by improving law, management, and policy.

Publications: The Environmental Law Reporter; National Wetlands Newsletter; The Environmental Forum.

Fish and Wildlife Information Exchange (FWIE)
2206 S. Main Street, Suite B
Blacksburg, VA 24060
Phone: (703) 231–7348
Fax: (703) 231–7019

The FWIE, founded in 1984, acts as a clearinghouse and technical assistance center to state and federal fish and wildlife agencies, helping them with databases and computer applications.

Publications: FWIE Newsletter.

Forest Farmers Association (FFA)
P.O. Box 95385
Atlanta, GA 30347
Phone: (404) 325–2954
Fax: (404) 325–2955

Founded in 1941, the FFA seeks to expand the timber industry and bring commercial forest areas of the 16 southern states to maximum production. The group also works to reduce waste caused by fire, insects, disease, and inefficient harvesting and processing. The FFA wants to develop a favorable tax base and encourage government and private forest and forest products research.

Publications: Forest Farmer magazine, bimonthly; *Forest Farmer Manual,* biennial.

Forest Trust (FT)
P.O. Box 519
Santa Fe, NM 87504–0519
Phone: (505) 983–8992
Fax: (505) 986–0798

The FT was founded in 1984 to improve the integrity, resilience, and productivity of forest and rangelands. The group is sponsored by the Tides Foundation, which is dedicated to the protection of America's forests.

Publications: Forest Trust Two-Year Report; Practitioner, quarterly newsletter.

Freshwater Foundation (FF)
Spring Hill Center
725 County Road 6
Wayzata, MN 55391
Phone: (612) 449–0092

Founded in 1986, the FF encourages proper use and management to keep surface water and groundwater usable for human consumption, industry, and recreation. The organization seeks to help people understand water issues and their environmental, political, social, and economic impact.

Publications: Facets of Freshwater, newsletter; *Health and Environment Digest; U.S. Water News;* special reports; Minnesota weather guide calendars.

Friends of the Earth (FOE)
218 D Street SE
Washington, DC 20003
Phone: (202) 544–2600
Fax: (202) 543–4710

FOE, founded in 1969, is devoted to influencing public policies and attitudes about issues such as ozone depletion, global warming, toxic chemical safety, coal mining, coastal and ocean pollution, tropical forest destruction, groundwater contamination, corporate accountability, and nuclear weapons production. FOE seeks to protect the planet, preserve biological, cultural, and ethnic diversity, and empower citizens to have a voice in decisions affecting the environment.

Publications: Atmosphere, quarterly newsletter; *Sparkling Hype at a Premium Price; Community Plume,* periodic newsletter; *Friends of the Earth/Not Man Apart* magazine, ten issues per year; *Groundwater News,* periodic newsletter; *Saving Our Skins: Technical Potential and Policies for the Elimination of Ozone-Depleting Chlorine Compounds.*

The Fund for Animals (FA)
200 W. 57th Street
New York, NY 10019
Phone: (212) 246–2096

Founded in 1967, the FA seeks to preserve wildlife, save endangered species, and promote the humane treatment of all animals. To this end, the group attempts to influence the government to fulfill its responsibilities to protect animals.

Gaia Institute (GI)
Cathedral of St. John the Divine
1047 Amsterdam Avenue at 112th Street
New York, NY 10025
Phone: (212) 295–1930
Fax: (212) 295–1930

The GI was founded in 1985 to promote the Gaia Hypothesis, which holds that conditions on Earth's surface are maintained and affected by the interactions among ecological communities. The group seeks to enhance understanding of global problems created by industrial societies, including acid rain, ozone depletion, and worldwide distribution of toxins.

Publications: The Gaia Newsletter, quarterly.

Game Conservation International (GCI)
P.O. Box 17444
San Antonio, TX 78217

Phone: (210) 824–7509
Fax: (210) 829–1355

GCI, founded in 1967, is dedicated to responsible, sustainable use of fish and wildlife and to preserving our hunting and fishing heritage for future generations.

Publications: Game Coin.

Global Coral Reef Alliance (GCRA)
324 N. Bedford Road
Chappaqua, NY 10514
Phone: (914) 238–8788

The GCRA conducts scientific research on coral reef ecosystems and the environmental factors affecting them, especially on the significance and spread of mass coral reef "bleaching," which is caused by a rise in ocean temperature and threatens the existence of the reefs.

Grassland Heritage Foundation (GHF)
P.O. Box 394
Shawnee Mission, KS 66201
Phone: (913) 677–3326

The GHF, founded in 1976, seeks to advance public under-standing and appreciation of the cultural, historical, and scientific value of native American grassland. The organization wants to acquire and preserve representative tracts of native prairie and produce educational materials for schools and the general public.

Publications: Bear News.

Greenpeace U.S.A. (GPUSA)
1436 U Street NW
Washington, DC 20009
Phone: (202) 462–1177
Fax: (202) 462–4507

Founded in 1979, GPUSA initiates active, though nonviolent, measures to aid endangered species, such as placing boats be-tween harpooners and whales and placing members between hunters and seal pups.

Publications: Greenpeace, quarterly.

Greensward Foundation (GF)
104 Prospect Park West
Brooklyn, NY 11215

The GF, founded in 1964, works for the improvement of natural-landscape urban parks, the understanding of these parks by the public, and the proper care of the parks by their custodians. The group encourages hiring and properly supervising unskilled and unemployed workers to maintain urban parks.

Publications: Central Park Map; Central Park/Prospect Park—A New Perspective; Little News, quarterly bulletin; *The Making of Prospect Park; The Men Who Made Central Park; Prospect Park/Brooklyn Botanic Garden Map; Prospect Park Handbook; Rock Trails in Central Park; Rockefeller New York; Tree Trails in Central Park; Tree Trails in Prospect Park.*

Hardwood Research Council (HRC)
P.O. Box 34518
Memphis, TN 38184–0518
Phone. (901) 377–1824
Fax: (901) 382–6419

The HRC was founded in 1953 to promote research on hardwood forest management and utilization.

Publications: Hardwood Forestry Bulletin; Proceedings: Annual Hardwood Symposium; technical reports.

Institute for Conservation Leadership (ICL)
2000 P Street NW, Suite 413
Washington, DC 20036
Phone: (202) 466–3330
Fax: (202) 659–3897

The ICL seeks to train and empower volunteer leaders and to build volunteer institutions that protect and conserve Earth's environment.

Institute for Earth Education (IEE)
Cedar Cove
Greenville, WV 24945
Phone: (304) 832–6404
Fax: (304) 832–6077

The IEE, founded in 1974, seeks to promote an understanding of, appreciation for, and harmony with Earth's natural systems and communities.

Publications: Talking Leaves Journal; Earth Education Sourcebook; Earth Education: A New Beginning.

**INTECOL—International Association
for Ecology (IAE)**
Drawer E
Aiken, SC 29802
Phone: (803) 725–2472
Fax: (803) 725–3309

Founded in 1967, the IAE promotes and communicates the science of ecology and the application of ecological principles to global needs. The organization collects and evaluates information, acts as the clearinghouse and center for coordination and dissemination of information and materials related to ecology and the global environment, coordinates application of ecological principles, and encourages public awareness of the economic and social importance of ecology.

Publications: Books: *Ecology International*, annual; *INTECOL Newsletter*, bimonthly; monographs.

**International Association of Fish
and Wildlife Agencies (IAFWA)**
444 N. Capitol Street NW, Suite 544
Washington, DC 20001
Phone: (202) 624–7890

The IAFWA is an association of states and territories of the United States, provinces of Canada, the Commonwealth of Puerto Rico, the United States government, the Dominion Government of Canada, and other governments of countries in the Western Hemisphere as well as individual associate members whose principal objective is the conservation, protection, and management of wildlife and related natural resources.

Publications: Newsletter; annual proceedings.

International Association of Wildland Fire (IAWF)
P.O. Box 328
Fairfield, WA 99012–0328

Phone: (800) 697–3443
Fax: (509) 283–2264

The IAWF, founded in 1983, acts as a forum for the exchange of information on wildland fires and fire prevention by fire scientists, managers, and individuals concerned with wildland fires.

Publications: Bibliography of Wildland Fire; Directory of Wildland Fire; International Bulletin of Wildland Fire, monthly; *International Journal of Wildland Fire.* Also distributes books and other materials on forest fire.

International Ecology Society (IES)
1471 Barclay Street
St. Paul, MN 55106–1405
Phone: (612) 774–4971

The IES was founded in 1976 to protect wildlife, domestic animals, and the environment through legislation, litigation, public action, and education.

Publications: Eco-Humane Letter, periodic newsletter; *Sunrise,* monthly newspaper.

The International Marine Mammal Project (IMMP)
Earth Island Institute
300 Broadway, Suite 28
San Francisco, CA 94133
Phone: (800) DOLPHIN, (800) 4-WHALES, (415) 788–3666

The IMMP, established in 1982, works toward an end to the slaughter of dolphins by the United States and international tuna industries.

Publications: Earth Island Journal; Dolphin Alert.

International Professional Hunters' Association (IPHA)
P.O. Box 17444
San Antonio, TX 78217
Phone: (210) 824–7506

Founded in 1969, the IPHA seeks to promote the good management of wildlife worldwide. The group collaborates with officers of game departments, national parks, and reserves in the conservation and management of fauna and flora and supports anti-poaching programs and game surveys. IPHA members are pledged to fair chase and sportsmanlike conduct in sport hunting and game photography.

**The International Society for the Preservation
of the Tropical Rainforest (ISPTR)**
3931 Camino de la Cumbre
Sherman Oaks, CA 91423
Phone: (818) 788–2002
Fax: (818) 990–3333

The ISPTR was organized in 1984 to promote park implementation, sustainable agriculture, and timber harvesting to help preserve tropical rain forests worldwide.

Publications: Tropical Rainforest: A World Survey of Our Most Valuable and Endangered Habitat; Amazon Hotline.

International Society of Tropical Foresters (ISTF)
5400 Grosvenor Lane
Bethesda, MD 20814
Phone: (301) 897–8720
Fax: (301) 897–3690

Founded in 1961, the ISTF seeks to enhance the protection, conservation, management, and wise use of tropical forests by technology transfer to natural resource managers and researchers in tropical forests.

Publications: ISTF News, quarterly, includes directory in the March edition; *ISTF Noticias* (in Spanish), quarterly; *ISTF Nouvelles* (in French), quarterly.

International Tree Crops Institute U.S.A. (ITCI)
P.O. Box 4460
Davis, CA 95617
Phone: (916) 753–4535
Fax: (916) 753–4535

The ITCI, founded in 1977, seeks to promote the scientific development of agroforestry (the growing of trees and shrubs for multiple uses, including food, fodder, fuel, and shelter). The organization seeks to increase the agricultural productivity of marginal lands and works to conserve soil, water, and energy resources, implement the use of windbreaks, and increase the use of arid and temperate-zone tree crops (carob, kiawe, honey locust, casuarina).

Publications: International Tree Crops, semiannual magazine; *Technical Papers*, periodic.

International Union for Conservation of Nature and Natural Resources (IUCN), World Conservation Union
Rue Mauverney 28
CH–1196 Gland, Switzerland
Phone: 022–999001
Telex: 419624 IUCN CH
Fax: 022–999-002
U.S. Office (IUCN U.S.)
1400 16th Street NW
Washington, DC 20036
Phone: (202) 797–5454
Fax: (202) 797–5461

The IUCN was founded in 1948 to promote scientifically based development that is sustainable and provides a lasting improvement in the quality of life for people all over the world.

Publications: IUCN Bulletin; Red Data Books (describing threatened species of mammals, amphibians, reptiles, invertebrates, and plants); *United Nations List of National Parks and Protected Areas; World Conservation Strategy; Living Resource Conservation for Sustainable Development; Caring for the Earth—A Strategy for Sustainable Living.*

International Whaling Commission (IWC)
The Red House
135 Station Road
Histon, Cambridge CB4 4NP
England
Phone: 0223233971

The IWC was established in 1946 under the International Convention for the Regulation of Whaling to provide for the conservation of whale stocks and the orderly development of the whaling industry. Member nations: Antigua and Barbuda, Argentina, Australia, Brazil, Chile, Costa Rica, Denmark, Dominica, Ecuador, Finland, France, Germany, India, Ireland, Japan, Kenya, Mexico, Monaco, Netherlands, New Zealand, Norway, Oman, People's Republic of China, Peru, Republic of Korea, Russian Federation, St. Kitts-Nevis, St. Lucia, St. Vincent and the Grenadines, Senegal, Seychelles, South Africa, Spain, Sweden,

Switzerland, the United Kingdom, the United States, and Venezuela.

Publications: Annual reports of the commission (including reports and papers of the scientific committee); special issues series on specialized cetacean subjects.

International Wild Waterfowl Association (IWWA)
Hidden Lake Waterfowl
5614 River Styx Road
Medina, OH 44256
Phone: (216) 725–8782

The IWWA, founded in 1958, seeks protection, conservation, and reproduction of any species of waterfowl that is considered in danger of eventual extinction. IWWA encourages the breeding of well-known and rare species in captivity.

Publications: First Breedings of Wild Waterfowl, periodic; *IWWA Membership Roster,* annual; *IWWA Newsletter,* quarterly; *Surplus Lists of Waterfowl,* semiannual.

International Wilderness Leadership Foundation (IWLF)
211 W. Magnolia
Fort Collins, CO 80521
(303) 498–0303

Organized in 1974, the IWLF seeks to protect wilderness and wildlife, to provide environmental experience and training, and to promote the use of wildlands worldwide.

Publications: The Leaf Newsletter; Wilderness Management; For the Conservation of Earth; Wilderness, the Way Ahead.

International Wildlife Coalition (IWC)
70 E. Falmouth Highway
East Falmouth, MA 02536
Phone: (508) 548–8328
Fax: (508) 548–8542

The IWC, founded in 1982, seeks to preserve wildlife and wildlife habitat in New England, Sri Lanka, Brazil, Australia, the United Kingdom, and Canada.

Publications: Whales of the World; Whalewatch, quarterly; *Wildlife and You; Wildlife Watch,* quarterly.

Izaak Walton League of America (IWLA)
1401 Wilson Boulevard, Level B
Arlington, VA 22209
Phone: (703) 528–1818
Fax: (703) 528–1836

Founded in 1922, the IWLA seeks to educate the public to conserve, maintain, protect, and restore the soil, forest, water, and other natural resources of the United States.

Publications: League Leader, bimonthly newsletter; *National Directory,* annual; *Outdoor America,* quarterly magazine; *Outdoor Ethics Newsletter,* quarterly; *Splash!* quarterly.

The Keystone Center (KC)
P.O. Box 8606
Keystone, CO 80435
Phone: (303) 468–5822

Founded in 1975, the KC acts as a center for environmental dispute resolution and education. The group conducts national policy dialogues on environmental, energy, natural resources, health, and science and technology issues. The KC operates a year-round residential environmental education center for children and adults on a campus at 9,300 feet in Old Keystone Village; the organization also provides an annual series of scientific symposia on molecular and cellular biology.

Publications: Consensus Newsletter; Discovery Newsletter.

Kids for Saving Earth (KSE)
620 Mendelssohn, Suite 145
Golden Valley, MN 55427
Phone: (612) 525–0002
Fax: (612) 525–0243

KSE was founded in 1989 as a grassroots organization that seeks to educate and empower children to unite with friends and take positive, peaceful action to protect Earth's environment.

Publications: Kids for Saving Earth News.

Land Improvement Contractors of America (LICA)
P.O. Box 9
1300 Maybrook Drive

Maywood, IL 60153
Phone: (708) 344–0700

LICA was founded in 1951 to foster efficient business principles in private and free enterprise in the fields of soil saving, soil building, and promoting clean water. LICA is a federation of associations of land improvement, excavation, and conservation contractors.

Publications: Applications Handbook; brochures; *LICA Membership Directory and Buyer's Guide,* annual; monthly newsletter; *Official Handbook.*

Legal Environmental Assistance Foundation (LEAF)
1115 N. Gadsden Street
Tallahassee, FL 32303–6327
Phone: (904) 681–2591
Fax: (904) 224–1275

LEAF was founded in 1979 to promote the protection of the environment and the health of the community by enforcing environmental regulations, discouraging harmful toxic and hazardous waste dumping, and encouraging energy efficiency.

Publications: LEAF Briefs, quarterly newsletter.

Migratory Bird Conservation Commission (MBCC)
Interior Building
Washington, DC 20240
Phone: (703) 358–1716

The MBCC, organized in 1929, considers, passes upon, and fixes the prices for lands recommended by the secretary of the interior for purchase or lease as migratory bird refuges in the National Wildlife Refuge System. The purchases are guided by the Migratory Bird Conservation Act of 18 February 1929 as amended.

National Arbor Day Foundation (NADF)
100 Arbor Avenue
Nebraska City, NE 68410
Phone: (402) 474–5655

The NADF, founded in 1971, sponsors Trees for America, Arbor Day, Tree City USA, Conservation Trees, the Arbor Day Institute, and Rain Forest Rescue educational programs.

Publications: Arbor Day; Tree City USA Bulletin; Conservation Trees, booklet; *Celebrate Arbor Day,* booklet; *Grow Your Own Tree; Trees Are Terrific.*

National Association of Environmental Professionals (NAEP)
5165 MacArthur Boulevard NW
Washington, DC 20016–3315
Phone: (202) 966–1500
Fax: (202) 966–1977

Founded in 1975, the NAEP promotes ethical practice in the environmental field and the recognition of the environmental profession as a distinct career. The organization maintains a professional certification program.

Publications: Environmental Professional, quarterly; bimonthly newsletter.

National Audubon Society (NAS)
700 Broadway
New York, NY 10003–9501
Phone: (212) 979–3000

The NAS, founded in 1905, seeks to protect the air, the water, the land, and the habitat critical to our health and the health of the planet via solid science, policy research, forceful lobbying, litigation, citizen action, and education.

Publications: Audubon; American Birds; Audubon Adventures; Audubon Activist.

National Coalition for Marine Conservation (NCMC)
5105 Paulsen Street, Suite 243
Savannah, GA 31405
Phone: (212) 354–0234

The NCMC was founded in 1973 to encourage the conservation of ocean fish and the protection of their environment.

Publications: Marine Bulletin; Ocean View; Marine Index; Currents.

National Council on Private Forests (NCPF)
P.O. Box 2000
Washington, DC 20013
Phone: (202) 667–3300

The NCPF, founded in 1985, informs decision makers in the private and public sectors of the effects of their policies on the management of private forest lands.

National Fish and Wildlife Foundation (NFWF)
1120 Connecticut Avenue NW, Suite 900
Washington, DC 20036
Phone: (202) 857–0166
Fax: (202) 857–0162

The NFWF, founded by Congress in 1984, leverages public and private funds by awarding challenge grants with private and federally appropriated funds for innovative and effective conservation activities.

National Hunters Association (NHA)
P.O. Box 820
Knightdale, NC 27545
Phone: (919) 365–7157

The NHA was founded in 1976 to protect hunting rights in the United States and around the world and to push for hunter safety. The group seeks the preservation of the right of the individual sportsperson to pursue the sport of hunting and the preservation of an adequate supply of game to hunt, both now and in the future.

Publications: NHA Newsletter.

National Institute for Urban Wildlife (NIUW)
10921 Trotting Ridge Way
Columbia, MD 21044
Phone: (301) 596–3311

The NIUW was founded in 1973 to conduct research on the relationship between humans and wildlife under urban and urbanizing conditions. The organization develops and disseminates practical procedures for maintaining, enhancing, and controlling wildlife species in urban areas.

Publications: Guide to Urban Wildlife Management; Integrating Man and Nature in the Metropolitan Environment; Urban Wildlife Manager's Notebook, quarterly; *Urban Wildlife News,* quarterly newsletter; *Wildlife Reserves and Corridors in the Urban Environment.*

National Park Foundation (NPF)
1101 17th Street NW
Washington, DC 20036
Phone: (202) 785–4500

The NPF was founded in 1967 and chartered by the U.S. Congress to provide private-sector support for the overall enhancement of the National Park System. The organization seeks to foster public awareness, appreciation, and understanding of the park system, as well as involvement in the system, through a focused set of independently driven and supplementary assistance programs.

Publications: The Complete Guide to America's National Parks.

National Park Trust (NPT)
P.O. Box 40236
Washington, DC 20016
Phone: (202) 625–2268

The NPT, established by the National Parks and Conservation Association in 1983, became an independent private nonprofit citizen land conservancy in 1990, saving parklands of national significance. The group works to acquire inholdings from willing sellers and hold them in trust until the National Park Service is able to include them in the national parks system.

National Parks and Conservation Association (NPCA)
1776 Massachusetts Avenue NW
Washington, DC 20036
Phone: (202) 223–6722
Fax: (202) 659–0650

The NPCA was founded in 1919 to preserve and expand our nation's park system. The organization has been an advocate as well as a constructive critic of the National Park Service. The NPCA has focused on the health of the entire system, from specific sites and programs to the processes of planning, management, and evaluation.

Publications: National Parks; ParkWatcher.

National Rifle Association of America (NRA)
1600 Rhode Island Avenue NW
Washington, DC 20036

Phone: (202) 828–6000
Fax: (202) 728–0340

The NRA was organized in 1871 to protect the right to possess and use arms and to train people in marksmanship and the safe handling and efficient use of small arms, as well as to foster and promote the shooting sports. The group seeks to promote hunter safety and to promote hunting as a sport and as a method of fostering the propagation, growth, conservation, and wise use of our renewable wildlife resources.

Publications: American Rifleman; American Hunter; InSights; Shooting Sports USA.

National Trappers Association (NTA)
207 W. Jefferson Street
P.O. Box 3667
Bloomington, IL 61702
Phone: (309) 829–2422

The NTA was founded in 1959 to promote sound conservation legislation, to conserve the nation's natural resources, to assist in implementing environmental education programs, and to promote a continued annual furbearer harvest as a necessary wildlife management tool.

Publications: American Trapper.

National Water Resources Association (NWRA)
3800 N. Fairfax Drive, Suite 4
Arlington, VA 22203
Phone: (703) 524–1544

The NWRA promotes the development, conservation, and management of water resources in the western states.

Publications: National Water Line; Water Writes.

National Wildflower Research Center (NWRC)
2600 FM 973 North
Austin, TX 78725
Phone: (512) 929–3600

The NWRC, founded in 1982, promotes the reestablishment and conservation of native plants by promoting their use in public and private landscape designs. The organization serves as a national clearinghouse of information on native plants, their sources, landscaping with them, and appropriate resource organizations and agencies. The NWRC conducts field research to learn more about the cultivation, propagation, and botanical properties of wildflowers and native plants.

Publications: Wildflower Newsletter; Wildflower Journal.

National Wildlife Federation (NWF)
1400 16th Street NW
Washington, DC 20036–2266
Phone: (202) 797–6800
Laurel Ridge Conservation Education Center
8925 Leesburg Pike
Vienna, VA 22184–0001
Phone: (703) 790–4000

The NWF was founded in 1936 to educate, inspire, and assist individuals and organizations of diverse cultures to conserve wildlife and other natural resources and to protect Earth's environment in order to achieve a peaceful, equitable, and sustainable future.

Publications: International Wildlife; National Wildlife; Ranger Rick; Your Big Backyard; EnviroAction; Conservation Directory; The Leader, newspaper; catalog of environmental education materials.

National Wildlife Refuge Association (NWRA)
10824 Fox Hunt Lane
Potomac, MD 20854

The NWRA was founded in 1975 to help protect the integrity of the National Wildlife Refuge System and to increase public understanding and appreciation of national wildlife refuges.

Publications: Blue Goose Flyer.

National Woodland Owners Association (NWOA)
374 Maple Avenue E, Suite 210
Vienna, VA 22180
Phone: (703) 255–2700

The NWOA was founded in 1983 to be a voice for owners of private nonindustrial woodland on forestry, wildlife, and resource conservation issues and to foster good stewardship of those forestlands.

Publications: Woodland Report, monthly newsletter; *National Woodlands* magazine, quarterly.

Natural Resources Council of America (NRCA)
801 Pennsylvania Avenue SE, Suite 410
Washington, DC 20003
Phone: (202) 547–7553

The NRCA, founded in 1946, provides member organizations with information on actions by Congress and the executive branch, makes available scientific data on conservation problems, and facilitates communication and cooperation among major national and regional organizations concerned with the sound management of natural resources in the public interest.

Natural Resources Defense Council (NRDC)

40 W. 20th Street
New York, NY 10011
Phone: (212) 727–2700

71 Stevenson Street
San Francisco, CA 94105
Phone: (415) 777–0220

1350 New York Avenue NW
Washington, DC 20005
Phone: (202) 783–7800

617 S. Olive Street
Los Angeles, CA
Phone: (213) 892-1500

212 Merchant Street, Suite 203
Honolulu, HI 96813
Phone: (808) 533–1075

Founded in 1970, the NRDC seeks to protect America's endangered natural resources and to improve the quality of the human environment, focusing on air and water pollution, global warming, nuclear safety, land use, the urban environment, toxic substances control, resource management, wilderness and wildlife protection, the global environment, energy conservation, agriculture, and forestry.

Publications: The Amicus Journal. A complete list of the NRDC's books and reports is available upon request.

The Nature Conservancy (NC)
1815 N. Lynn Street
Arlington, VA 22209
Phone: (703) 841–5300
Fax: (703) 841–1283

The NC, founded in 1951, seeks to preserve biological diversity by protecting natural lands and the life they harbor. The organization manages a system of over 1,600 sanctuaries nationwide and identifies ecologically significant natural areas by working with "natural heritage programs" in each state.

Publications: Nature Conservancy magazine.

North American Wildlife Foundation (NAWF)
102 Wilmot Road, Suite 410
Deerfield, IL 60015
Phone: (708) 940–7776

Founded in 1911, the NAWF supports effective, practical, and systematic research on management practices and techniques to benefit wildlife and other natural resources in the public interest. The group owns the Delta Waterfowl Research Station in Manitoba, Canada.

North American Wildlife Park Foundation (NAWPF)
Wolf Park
Battle Ground, IN 47920
Phone: (317) 567-2265

The NAWPF, founded in 1972, operates Wolf Park for the study of wolf behavior.

Publications: Wolf Park News.

Organization of Wildlife Planners (OWP)
Utah Division of Wildlife Resources
1596 W. North Temple
Salt Lake City, UT 84116
Phone: (801) 538–4706
Fax: (406) 444–4952

The OWP, founded in 1979, is dedicated to developing an organized, objective-oriented approach to the management of North America's fish, wildlife, and natural resources.

Publications: Comprehensive Planning for Wildlife Resources; Proceedings of the OWP Conference, annual, includes membership list; *Tomorrow's Management,* quarterly newsletter.

Outdoor Writers Association of America (OWAA)
2017 Cato Avenue, Suite 101
State College, PA 16801
Phone: (814) 234–1011

Founded in 1927, the OWAA strives to increase outdoor writers' knowledge and understanding of the conservation of natural resources.

Publications: Outdoors Unlimited.

Rachel Carson Council (RCC)
8940 Jones Mill Road
Chevy Chase, MD 20815
Phone: (301) 652–1877

The RCC, founded in 1965, seeks to inform both scientists and laypeople about chemical contamination of the environment, especially by pesticides.

Publications: Basic Guide to Pesticides.

Rainforest Action Network (RAN)
450 Sansome, Suite 700
San Francisco, CA 94111
Phone: (415) 398–4404
Fax: (415) 398–2732

Founded in 1985, RAN seeks to protect rain forests using direct action, including letter-writing campaigns, boycotts, and demonstrations against corporations and lending agencies contributing to rain forest destruction.

Publications: World Rainforest Report; Action Alert.

Rainforest Alliance (RA)
65 Bleecker Street
New York, NY 10012
Phone: (212) 677–1900
Fax: (212) 677–2187

The RA, founded in 1986, seeks to develop and promote economically viable and socially desirable alternatives to deforestation as a means to protect the world's endangered tropical forests.

Publications: The Canopy.

Rare Center for Tropical Conservation (RCTC)
1529 Walnut Street
Philadelphia, PA 19102

Phone: (215) 568–0420
Fax: (215) 568–0516

Founded in 1973, the RCTC seeks to preserve threatened habitats and ecosystems in Latin America and the Caribbean by monitoring endangered birds as environmental indicators.

Save-the-Redwoods League (SRL)
114 Sansome Street, Room 605
San Francisco, CA 94104
Phone: (415) 362–2352

The SRL was founded in 1918 to help save representative areas of our primeval forests, especially redwood parks.

Publications: California Redwood Parks and Preserves; Trees, Shrubs and Flowers of the Redwood Region; Redwoods of the Past; Story Told by a Fallen Redwood.

Scenic America (SA)
21 Dupont Circle NW
Washington, DC 20036
Phone: (202) 833–4300

SA, founded in 1981, provides information and technical assistance on scenic byways, tree preservation, the economics of aesthetic regulations, billboard and sign control, scenic area preservation, growth management, and all forms of aesthetic regulation.

Publications: Viewpoints, newsletter; *Economics of Community Character Preservation.*

Sierra Club (SC)
730 Polk Street
San Francisco, CA 94109
Phone: (415) 776–2211

Founded in 1892, the SC strives to promote the responsible use of Earth's ecosystems and resources and to educate and enlist humanity to protect and restore the quality of the natural and human environment.

Publications: Sierra; Grassroots Sierra.

Smithsonian Institution (SI)
1000 Jefferson Drive SW

Washington, DC 20560
Phone: (202) 357–2700

The SI was founded in 1846 as an independent trust instrument
of the United States to "increase and diffuse knowledge" on arts,
sciences, and history.

Publications: Smithsonian magazine.

Society for Conservation Biology (SCB)
Department of Biology
University of Nevada
Reno, NV 89557–0015
Phone: (702) 784–6188

Founded in 1985, the SCB is a professional society dedicated to
providing the scientific information and expertise required to
protect the world's biological diversity.

Publications: Conservation Biology.

Society for Ecological Restoration (SER)
University of Wisconsin–Madison Arboretum
1207 Seminole Highway
Madison, WI 53711
Phone: (608) 262–9547
Fax: (608) 262–9547

The SER, founded in 1987, seeks to promote the development of
ecological restoration and to raise awareness about the value and
limitations of restoration as a conservation strategy.

Publications: Restoration and Management Notes; SER News; Restoration Ecology; Ecological Restoration: The New Management Challenge; Proceedings of the First SER Conference, 1989.

Society of American Foresters (SAF)
5400 Grosvenor Lane
Bethesda, MD 20814
Phone: (301) 897–8720
Fax: (301) 897–3690

The SAF was founded in 1900 to advance the science, technology,
education, and practice of professional forestry and to use the
knowledge and skills of the profession to benefit society. The or-

ganization is also an accreditation authority for professional forestry education in the United States.

Publications: Journal of Forestry, monthly; *Forest Science,* quarterly; *Southern Journal of Applied Forestry,* quarterly; *Northern Journal of Applied Forestry,* quarterly; *Western Journal of Applied Forestry,* quarterly.

Tall Timbers Research (TTR)
Route 1, Box 678
Tallahassee, FL 32312–9712
Phone: (904) 893–4153

TTR was founded in 1958 to study fire's natural role in ecosystems, to promote prescribed burning in land management, and to promote ecology, conservation, and management of native species.

Tread Lightly! (TL)
298 24th Street, Suite 325C
Ogden, UT 84401
Phone: (801) 627–0077

TL was organized in 1990 to promote responsible use of off-highway vehicles and other forms of backcountry travel and to work for the protection of public and private lands based on low-impact principles applicable to all recreational activities.

Publications: Tread Lightly! quarterly newsletter.

TreePeople (TP)
12601 Mulholland Drive
Beverly Hills, CA 90210
Phone: (818) 753–4600

TP was founded in 1973 to promote personal involvement, community action, and global environmental awareness. To help individuals recognize the power they have to change and improve the environment, the organization teaches individuals and groups how to plant and maintain trees, emphasizing the important role of diverse groups working together for the common good.

Publications: Seedling News; A Planter's Guide to the Urban Forest; TreePeople, newsletter; *The Simple Act of Planting a Tree.*

Trust for Public Land (TPL)
116 New Montgomery Street, Fourth Floor
San Francisco, CA 94105
Phone: (415) 495–4014
Fax: (415) 495–4103

The TPL was founded in 1972 to train and provide technical assistance to urban and community groups on how to acquire and preserve land in urban and rural areas for public use. The organization seeks to acquire recreational, historic, and scenic lands for conveyance to local, state, and federal agencies and nonprofit organizations.

Publications: Annual report; *Land and People,* semiannual magazine; semiannual newsletter.

Western Ancient Forest Campaign (WAFC)
1400 16th Street NW, Suite 294
Washington, DC 20036
Phone: (202) 939–3324

The WAFC, founded in 1991, strives to inform the general public and policy makers about the inherent environmental value of ancient forest ecosystems and to reform federal forest management policies. The group sponsors Ancient Forest Roadshows and the Western Ancient Forest Expedition, an ancient forest log traveling exhibit.

Publications: East Side Ancient Forest, brochure; pamphlets; *Report from Washington,* biweekly.

Wetlands for the Americas (WA)
81 Stage Point Road
P.O. Box 1770
Manomet, MA 02345
Phone: (508) 224–6521
Fax: (508) 224–9220

WA was formerly the Western Hemisphere Shorebird Reserve Network (WHSRN). WA seeks to implement conservation activities that advance the health of wetland ecosystems and species. WHSRN continues to function as a program promoting international cooperation in migratory species conservation.

Publications: Wetlands for the Americas.

Wetlands for Wildlife (WW)
P.O. Box 344
West Bend, WI 53095

WW, founded in 1960, acquires wetlands and wildlife habitat in the United States that is transferred to federal, state, and county agencies to be maintained and managed exclusively for public purposes.

Publications: Wetlands for Wildlife, newsletter.

Wilderness Education Association (WEA)
20 Winona Avenue, Box 89
Saranac Lake, NY 12983
Phone: (518) 891–2915, ext. 254
Fax: (518) 891–2915, ext. 214.

Founded in 1978, the WEA offers wilderness leadership training in the hopes of improving the safety and quality of outdoor trips and enhancing the conservation of the wild outdoors.

Publications: WEA Legend; Trustees and Affiliates Briefing System.

The Wilderness Society (WS)
900 17th Street NW
Washington, DC 20006–2596
Phone: (202) 833–2300

The WS was founded in 1935 to foster an American land ethic that seeks to preserve wilderness and wildlife, protecting America's prime forests, parks, rivers, and shores.

Publications: Wilderness.

Wilderness Watch (WW)
P.O. Box 782
Sturgeon Bay, WI 54235
Phone: (414) 743–1238

Founded in 1969, WW promotes the sustained use of America's sylvan lands and waters. The group provides access to a scientific advisory staff of experts in the behavioral and physical sciences to help make sound decisions about ecological matters.

Publications: Watch It.

Wildfowl Trust of North America (WTNA)
P.O. Box 519
Grasonville, MD 21638
Phone: (410) 827–6694
Fax: (410) 827–6713

The WTNA was founded in 1979 to create and maintain wildfowl and wetlands centers where the best examples of conservation, education, and research would be available. The group organizes captive breeding programs for endangered species and releases native waterfowl species to the wild in an effort to restore historic regional populations.

Publications: Semiannual newsletter, includes calendar of events, children's page, and waterfowl and conservationist profiles.

Wildlife Conservation Fund of America (WCFA) and Wildlife Legislative Fund of America (WLFA)
801 Kingsmill Parkway
Columbus, OH 43229–1137
Phone: (614) 888–4868
Fax: (202) 888-0326

National Affairs Office
Washington, DC
Phone: (202) 466-4077
Fax: (202) 466-8727

These companion organizations were founded in 1978 to protect Americans' right to hunt, fish, and trap and to encourage scientific wildlife management practices.

Publications: A Lawyer's Primer; Lobbying; Protect What's Right, quarterly newsletter.

Wildlife Forever (WF)
12301 Whitewater Drive, Suite 210
Minnetonka, MN 55343
Phone: (612) 936–0605

WF was founded in 1987 to educate the public about the need for scientific wildlife management, to conserve wildlife and preserve wildlife habitat, and to fund special projects and studies of wildlife conservation.

Publications: Cry of the Wild; Wildlife Forever Update.

Wildlife Habitat Enhancement Council (WHEC)
1010 Wayne Avenue, Suite 920
Silver Spring, MD 20910
Phone: (301) 588–8994
Fax: (301) 588–4629

The WHEC was founded in 1987 to assist corporations in enhancing their undeveloped lands for the benefit of wildlife, fish, and plant life. The organization provides technical assistance in establishing and maintaining responsible corporate wildlife management practices.

Publications: WIN—Wildlife in the News; Economic Benefits of Wildlife Habitat Enhancement on Corporate Lands; Habitat Management Series; International Registry of Certified Corporate Wildlife Habitats.

Wildlife Information Center (WIC)
629 Green Street
Allentown, PA 18102
Phone: (215) 434 1637

WIC, founded in 1986, sponsors the Kittatinny Raptor Corridor Project and long-term hawk migration field studies at Bake Oven Knob, Pennsylvania, and secures and disseminates wildlife conservation information

Publications: Educational Hawkwatcher; Wildlife Book Review; Wildlife Conservation Reports; Wildlife Activist; American Hawkwatcher.

Wildlife Management Institute (WMI)
1101 14th Street NW, Suite 801
Washington, DC 20005
Phone: (202) 371–1808
Fax: (202) 408–5059

The WMI, founded in 1911 by concerned sportsmen/businessmen, seeks to improve professional management of natural resources for the benefit of those resources and North America, including its people. The WMI supports wise uses of wildlife, including regulated recreational hunting.

Publications: Outdoor News Bulletin; Transactions North American Wildlife and Natural Resources Conference; books and booklets.

Wildlife Preservation Trust International (WPTI)
3400 W. Girard Avenue
Philadelphia, PA 19104
Phone: (215) 222–3636
Fax: (215) 222–2191

The WPTI, founded in 1971, promotes captive breeding of endangered species to save them from extinction and encourages the reintroduction to the wild of captive-bred animals.

Publications: Annual report; *The Dodo,* annual journal; *Dodo Dispatch,* three issues per year, newsletter; *On the Edge,* three issues per year, newsletter.

The Wildlife Society (WS)
5410 Grosvenor Lane
Bethesda, MD 20814
Phone: (301) 897–9770

Founded in 1937, the WS promotes the sound stewardship of wildlife resources and the environments upon which wildlife and humans depend. The organization takes an active role in preventing human-induced environmental degradation, increases awareness and appreciation of wildlife values, and seeks the highest standards in all activities of the wildlife professions.

Publications: Journal of Wildlife Management; Wildlife Monographs; Wildlife Society Bulletin; The Wildlifer.

World Wildlife Fund (WWF)
1250 24th Street NW
Washington, DC 20037
Phone: (202) 293–4800

The WWF, founded in 1961, seeks to protect endangered wildlife and wildlands, especially in the tropical forests of Latin America, Asia, and Africa, via the creation and protection of national parks and nature reserves. The group monitors international trade in wildlife and promotes ecologically sound development to promote conservation of Earth's living resources.

Publications: Focus.

Organizations Concerned with Specific Species or Groups of Organisms

American Cetacean Society (ACS)
P.O. Box 2639
San Pedro, CA 90731–0943
Phone: (310) 548–6279
Fax: (310) 548–6950

The ACS, a nonprofit organization founded in 1967, works in the areas of conservation, education, and research to protect marine mammals—especially whales, dolphins, and porpoises—and the oceans they live in.

Publications: Whalewatcher; Journal of the American Cetacean Society; WhaleNews; ACS National Newsletter.

American Horse Protection Association (AHPA)
1000 29th Street NW, T100
Washington, DC 20007
Phone: (202) 965–0500

The AHPA was founded in 1966 to promote the protection and welfare of horses, both wild and domestic. The organization helped gain passage of the Horse Protection Act of 1970 and the Wild Horse and Burro Protection Act of 1971.

Publications: AHPA Newsletter, quarterly.

American Pheasant and Waterfowl Society (AP&WS)
W2270 U.S. Highway 10
Granton, WI 54436
Phone: (715) 238–7291

The AP&WS was founded in 1936 to help perpetuate all varieties of upland game, ornamental birds, and waterfowl.

Publications: APWS Magazine, ten issues per year; *Membership Roster,* periodic.

Atlantic Salmon Federation (ASF)
P.O. Box 429
St. Andrews, NB, Canada E0G 2X0
Phone: (506) 529–4581
Fax: (506) 529–4438

The ASF, founded in 1982, promotes the preservation and management of Atlantic salmon stocks. The group planned and built the North American Salmon Research Center to conduct research in salmon genetics.

Publications: Atlantic Salmon Journal, quarterly newsletter; *On the Rise,* quarterly newsletter; *SALAR,* quarterly newsletter; *Special Publication Series,* periodic.

Atlantic Waterfowl Council (AWC)
Division of Fish and Wildlife
P.O. Box 1401
Dover, DE 19903
Phone: (302) 739–5295
Fax: (302) 739–6157

Founded in 1952, the AWC coordinates waterfowl research and management in the Atlantic Flyway.

Publications: Techniques of Waterfowl Habitat Development and Management.

Bat Conservation International (BCI)
P.O. Box 162603
Austin, TX 78716–2603
Phone: (512) 327–9721
Fax: (512) 327–9724

The BCI, founded in 1982, promotes bat conservation and management efforts and increases public awareness of the value and ecological importance of bats, as well as the conservation needs of bats.

Publications: America's Neighborhood Bats; The Bat House Builder Handbook; Bats, quarterly; *Bats: A Creativity Book for Young Conservationists; Bats, Pesticides and Politics; Educators Activity Book.*

Beaver Defenders (BD)
Unexpected Wildlife Refuge
P.O. Box 765
Newfield, NJ 08344
Phone: (609) 697–3541

BD was founded in 1970 to preserve and protect beavers.

Publications: The Beaver Defenders, quarterly.

Billfish Foundation (BF)
2419 E. Commercial Boulevard, Suite 303
Fort Lauderdale, FL 33308
Phone: (305) 938–0150
Fax: (305) 938–5311

The BF was founded in 1986 to promote the conservation of billfish through scientific research and education. The organization is working to develop an international management plan to ensure the survival of billfish.

Publications: TBF Newsletter, quarterly.

Birds of Prey Rehabilitation Foundation (BPRF)
Rural Route 2, Box 659
Broomfield, CO 80020
Phone: (303) 460–0674

The BPRF was founded in 1984 to provide for the preservation and rehabilitation of birds of prey.

Publications: Brochures; pamphlets; *The Windwalker,* annual newsletter.

Canvasback Society (CS)
P.O. Box 101
Gates Mills, OH 44040
Phone: (216) 443–2340

The CS was founded in 1975 to promote research on the canvasback duck and to encourage the creation, restoration, and preservation of duck habitat

Cetacean Society International (CSI)
P.O. Box 343
Plainville, CT 06062–0343
Phone: (203) 793–8400

Founded in 1974, the CSI promotes the elimination of outlaw whaling and trading in whale products and strives to preserve and conserve cetaceans, including whales, dolphins, and porpoises, and provide legal protection for small cetaceans. The group works to establish whale sanctuaries, assists in rescuing stranded whales, and works to limit drift-net fishing.

Publications: Connecticut Whale, bimonthly newsletter.

Committee for Conservation and Care of Chimpanzees (CCCC)
3819 48th Street NW
Washington, DC 20016
Phone: (202) 362–1993
Fax: (202) 686–3402

The CCCC was founded in 1986 to assess population status, monitor trade, and provide guidelines for captive care of chimpanzees.

Committee for the Preservation of the Tule Elk (CPTE)
43 Keystone Way
San Francisco, CA 94127
Phone: (415) 333–7228

The CPTE was founded in 1960 to protect and preserve the Tule Elk and other rare species and to promote the aesthetic enjoyment of nature.

Publications: History of the California Tule Elk; Owens Valley, Home of the Tule Elk; The Return of the Tule Elk; Sanctuaries for the Protection of Rare Species.

Deer Unlimited of America (DUA)
P.O. Box 1129
Abbeville, SC 29620
Phone: (803) 391–2300

DUA, founded in 1977, advocates for the rights of sport hunters and provides a place for members to hunt by leasing land for bow and gun hunting.

Publications: Deer Unlimited, bimonthly magazine.

Desert Tortoise Council (DTC)
P.O. Box 1738
Palm Desert, CA 92261–1738
Phone: (619) 341–8449

Founded in 1975, the DTC seeks to ensure the survival of viable populations of the desert tortoise (*Xerobates agassizii*) throughout its existing range via studies of its life history, biology, and physiology and via management.

Publications: Annotated Bibliography of the Desert Tortoise; papers; *Proceedings of Symposium,* annual.

Digit Fund (DF)
45 Inverness Drive E, Suite B
Englewood, CO 80112
Phone: (303) 790–2349
Fax: (303) 790–4066

The DF was founded in 1978 by primatologist Dian Fossey (1932–1985) and was named in memory of a gorilla she named Digit, who was killed by poachers. The organization promotes the study of the endangered mountain gorillas of central Africa and seeks to protect them via preservation and conservation of the gorillas' rain forest habitat. The DF supports the Karisoke research center for the study of gorilla behavior and the environment and operates Rwandan antipoaching patrols to guard the gorillas and monitor their well-being.

Publications: Digit News, quarterly newsletter.

Ducks Unlimited (DU)
1 Waterfowl Way
Memphis, TN 38120–2351
Phone: (901) 758 3825
Fax: (901) 758–3850

DU, founded in 1937, seeks to fulfill the annual life cycle needs of North American waterfowl by protecting, enhancing, restoring, and managing important wetlands and associated uplands.

Publications: Ducks Unlimited magazine; *Puddler* magazine.

Elephant Research Foundation (EIG)
106 E. Hickory Grove
Bloomfield Hills, MI 48304
Phone: (810) 540–3947
Fax: (810) 540–3948

The EIG was founded in 1977 to promote public interest in and public knowledge of elephants, as well as to protect and conserve elephant species.

Publications: Bibliographies; *Elephant,* periodic journal, includes membership roster.

Foundation for North American Wild Sheep (FNAWS)
720 Allen Avenue
Cody, WY 82414–3402

Phone: (307) 527–6261
Fax: (307) 527–7117

The FNAWS, founded in 1977, seeks to prevent the extinction of all species of wild sheep native to North America by performing biological studies, acquiring buffer land, preventing poaching, and reestablishing wild sheep populations in historic habitats via the transplantation of sheep.

Publications: Wild Sheep, periodic.

Foundation for the Preservation and Protection of the Przewalski Horse (FPPPH)
Animal Dairy Science Department
Livestock Poultry Building
Athens, GA 30605
Phone: (706) 542–0976

The FPPPH, founded in 1977, strives to preserve and protect the Przewalski horse, which may be the only wild horse still in existence. Populations in captivity have a low breeding rate and die young due to high inbreeding.

Friends of the Australian Koala Foundation (FAKF)
50 W. 29th Street, Suite 9W
New York, NY 10001
Phone: (212) 779–0700
Fax: (212) 689–0376

FAKF was founded in 1989 to minimize the destruction of koala habitat and improve population dynamics. The group encourages biological research, disease control, and studies of habitat usage and preservation.

Publications: Australian Koala Foundation Newsletter, bimonthly.

Friends of the Sea Lion Marine Mammal Center (FSLMMC)

FSLMMC was founded in 1971 to aid sick and injured marine animals and then return them to freedom. The organization conducts citizen training on how to help beached marine animals and does basic research on the treatment of beached animals.

Great Bear Foundation (GBF)
P.O. Box 2699

Missoula, MT 59806
Phone: (406) 721–3009

The GBF was founded in 1982 to conserve the eight species of bears in the world. The organization operates a program that reimburses ranchers in select areas near the Rocky Mountains for livestock killed by grizzly bears.

Publications: Bear News, quarterly

The Hawk and Owl Trust (HOT)
Birds of Prey Section
London Zoo
Regent's Park
London NW1 4RY, England

HOT was founded in 1969 to conserve birds of prey, including owls, and encourage the appreciation of them.

Publications: Annual report; *Raptor Newsletters/Peregrine.*

Hawk Migration Association of North America (HMANA)

HMANA was founded in 1974 to advance the knowledge of bird-of-prey migration across the continent, to monitor raptor populations as an indicator of environmental health, to study further the behavior of raptors, and to contribute to greater public understanding of birds of prey.

Publications: Hawk Migration Studies; conference proceedings.

International Association for Bear Research and Management (IBA)
ADF&G
333 Raspberry Road
Anchorage, AK 99518–1599
Phone: (907) 344–0541
Fax: (907) 344–7914

The IBA, founded in 1968, seeks to foster communication and cooperation among research biologists and animal and land managers on management of bears and their habitat.

Publications: Bears—Their Biology and Management, triennial; proceedings; *International Bear News,* two to four issues per year. Newsletter.

International Crane Foundation (ICF)
11376 Shady Lane Road
Baraboo, WI 53913–9778
Phone: (608) 356–9462
Fax: (608) 356–9465

Founded in 1973, the ICF seeks to preserve and restock the crane population in its natural habitat and to increase captive propagation.

Publications: Crane Research around the World; Cranes, Cranes, Cranes; ICF Bugle, quarterly; *Proceedings of the 1983 International Crane Workshop; Reflections—A Study of Cranes.*

International Primate Protection League (IPPL)
P.O. Box 766
Summerville, SC 29484
Phone: (803) 871–2280
Fax: (803) 871–7988

The IPPL, founded in 1974, is devoted to the conservation and protection of nonhuman primates.

Publications: International Primate Protection League News.

International Snow Leopard Trust (ISLT)
4649 Sunnyside Avenue N, Suite 325
Seattle, WA 98103
Phone: (206) 632–2421
Fax: (206) 632–3967

The ISLT was founded in 1981 to conserve the snow leopard and its mountain habitat.

Publications: Bibliography; proceedings; *Snow Line,* periodic newsletter. Also publishes status review.

International Society for the Protection of Mustangs and Burros (ISPMB)
6212 E. Sweetwater Avenue
Scottsdale, AZ 85254
Phone: (602) 991–0273
Fax: (602) 991–0273

The ISPMB was founded in 1960 to protect wild horses and burros by preserving their habitat and preventing molestation or interference with the animals. The group sponsors a program for the placement and care of unadoptable wild horses and burros.

Publications: Wild Horse and Burro Diary, quarterly.

Mountain Lion Foundation (MLF)
P.O. Box 1896
Sacramento, CA 95812
Phone: (916) 442–2666

Founded in 1986, the MLF strives to preserve the cougar and enhance its long-term survival.

Publications: Cougar News, semiannual; *Cougar: The American Lion; Habitat Preservation; Preserving Cougar Country.*

National Foundation to Protect America's Eagles (NFPAE)
P.O. Box 120206
Nashville, TN 37212
Phone: (800) 2-EAGLES, (615) 847–4171

The NFPAE, founded in 1985, is dedicated to saving, restoring, and protecting America's endangered national symbol, the bald eagle.

Publications: American Eagle News.

National Wild Turkey Federation (NWTF)
P.O. Box 530
770 Augusta Road
Edgefield, SC 29824
Phone: (803) 637–3106
Fax: (803) 637–0034

The NWTF was founded in 1973 to conserve and protect American wild turkey populations as a valuable natural resource. The group maintains the Wild Turkey Center and sponsors an annual wild turkey stamp and print art program.

Publications: The Caller, quarterly newspaper; *Turkey Call,* bimonthly magazine.

North American Bear Society (NABS)
3875 N. 44th Street, Suite 102
Phoenix, AZ 85018
Phone: (602) 952–1810
Fax: (602) 952–8230

The NABS was founded in 1986 to promote the conservation and management of bears and other North American wildlife.

Publications: Bear Tracker, quarterly; *Ursus,* annual magazine.

North American Bluebird Society (NABS)
P.O. Box 6295
Silver Spring, MD 20916–6295
Phone: (301) 384–2798

Founded in 1978, the NABS seeks to increase interest in and conservation of bluebirds and other native cavity-nesting birds of North America.

Publications: Bluebird Bibliography; Sialia, quarterly journal.

North American Crane Working Group (NACWG)
2550 N. Diers Avenue, Suite H
Grand Island, NE 68803
Phone: (308) 384–4633
Fax: (308) 384–4634

The NACWG was founded in 1988 to promote the conservation of cranes and their habitat in North America.

Publications: Proceedings of the North American Crane Workshop, periodic; *Unison Call,* semiannual newsletter.

North American Loon Fund (NALF)
6 Lily Pond Road
Gilford, NH 03246
Phone: (603) 528–4711

The NALF was founded in 1979 to protect loons and their habitat and to construct artificial nesting sites. The group discourages lakeshore development in areas frequented by loons.

Publications: The Loon Call, newsletter, three issues per year; *Voices of the Loon.*

North American Wolf Society (NAWS)
P.O. Box 82950
Fairbanks, AK 99708
Phone: (907) 474–3755

The NAWS was founded in 1973 to encourage a rational approach to the conservation of the wolf and other wild canids of North America.

Publications: A List of Coyote Literature; A List of Wolf Literature; A Natural History of Coyotes; A Natural History of Wolves; Activities for Educators; Endangered Species Act; Predator Controls and Coyotes; Red Wolves; Wolf Hybrids; Wolf Recovery; Wolves as Pets; Wolves in Alaska.

Pacific Seabird Group (PSG)
Savannah River Ecology Lab
P.O. Drawer E
Aiken, SC 29801
Phone: (803) 725–2475
Fax: (803) 725–3309

The PSG was founded in 1972 to study and conserve Pacific seabirds and their marine environment.

Publications: Membership directory, periodic; *Pacific Seabird Group Bulletin,* semiannual.

Pacific Whale Foundation (PWF)
Kealia Beach Plaza, Suite 25
101 N. Kihei Road
Kihei, HI 96753
Phone: (808) 879–8811
Fax: (808) 879–2615

The PWF was founded in 1980 on the premise that whales, dolphins, and porpoises are biologically essential to Earth's ecosystem and must be conserved and protected.

Publications: Fin and Fluke Report, quarterly journal; monographs; *Soundings,* semiannual newsletter; *Whale One Dispatch,* periodic newsletter; *Whalewatching Guide.*

Pelican Man's Bird Sanctuary (PMBS)
1708 Ken Thompson Parkway
Sarasota, FL 34236
Phone: (813) 388–4444
Fax: (813) 388–3258

The PMBS was founded in 1985 to operate a rehabilitation center for injured pelicans, blue herons, gulls, and other birds and to promote wildlife protection and rescue.

Publications: Brochures; *The Peligram,* newsletter, three to four issues per year.

Peregrine Fund (PF)
World Center for Birds of Prey
5666 W. Flying Hawk Lane
Boise, ID 83709
Phone: (208) 362–3716
Fax: (208) 362–2376

Founded in 1970, the PF operates the World Center for Birds of Prey as a research and educational facility and maintains and propagates rare birds of prey in captivity. The group seeks to reestablish natural populations of raptors and conserve their habitat.

Publications: Annual report; brochures; *Falcon Propagation: A Manual on Captive Breeding; Hacking: A Method for Releasing Peregrine Falcons and Other Birds of Prey; Peregrine Falcon Populations: Their Management and Recovery; Wise as an Owl: A Resource and Teacher's Guide to Birds of Prey.*

Pheasants Forever (PF)
P.O. Box 75473
St. Paul, MN 55175
Phone: (612) 481–7142

PF was founded in 1982 to promote pheasant habitat conservation and rebuild lost wetlands, as well as reestablish shelterbelts (barriers of trees or shrubs that reduce erosion and provide protection for wildlife).

Publications: Annual report; *P.F. Flyer,* periodic newsletter.

Project Wolf U.S.A. (PWUSA)
168 Galer
Seattle, WA 98109
Phone: (206) 283–1957

PWUSA seeks to protect wolf populations in North America by changing present U.S. and Canadian game laws, under which wolves can legally be hunted and trapped in British Columbia and Alaska.

Purple Martin Conservation Association (PMCA)
Edinboro University of Pennsylvania
Edinboro, PA 16444
Phone: (814) 734–4420
Fax: (814) 734–5803

The PMCA was founded in 1987 to provide information on techniques of attracting purple martins and managing nesting sites.

Publications: Brochure; *Purple Martin Update,* quarterly magazine.

The Raptor Center (RC)
University of Minnesota
1920 Fitch Avenue
St. Paul, MN 55108

The RC, founded in 1974, strives to preserve biological diversity among raptors and other avian species through medical treatment, scientific investigation, education, and management of wild populations.

Publications: The Raptor Release.

Raptor Education Foundation (REF)
21901 E. Hampden Avenue
Aurora, CO 80013
Phone: (303) 680–8500

The REF, founded in 1980, uses nonreleasable raptors in a series of lecture programs illustrating the importance of raptors in the balance of nature.

Publications: Talon; Talon Supplement; Eagles, Hawks, Falcons, and Owls of America, a coloring album; *Castings,* volunteer newsletter.

Save the Manatee Club (SMC)
500 N. Maitland Avenue
Maitland, FL 32751
Phone: (407) 539–0990
Fax: (407) 539–0871

The SMC was founded in 1981 to save and protect the West Indian manatee, an endangered marine mammal, and its habitat by heightening public awareness and education regarding sanctuary areas, boating speed signs, and laws pertaining to protection of the manatees.

Publications: Adopt-a-Manatee, brochure; *Guidelines for Protecting Manatees; Manatees: An Educator's Guide; Save the Manatee Club,* brochure; *Save the Manatee Club Newsletter,* five issues per year; *West Indian Manatee,* booklet.

Save the Whales (SW)
P.O. Box 3650, Georgetown Station
Washington, DC 20007
Phone: (202) 337–2332
Fax: (202) 338–9478

SW was founded in 1971 to save great whales from extinction and to work for whale regeneration.

Publications: Brochures; *Whales versus Whalers;* posters.

Society for the Conservation of Bighorn Sheep (SCBS)
P.O. Box 801
La Canada, CA 91012
Phone: (818) 393–0706
Fax: (818) 393–1173

The SCBW was founded in 1963 to study and conserve bighorn sheep.

Publications: Sheepherder, quarterly newsletter.

The Trumpeter Swan Society (TTSS)
3800 County Road 24
Maple Plain, MN 55359
Phone: (612) 476–4663
Fax: (612) 476–1514

TTSS was founded in 1968 to promote research into the ecology and management of the trumpeter swan throughout North America and to advance the science and art of trumpeter swan management, both in captivity and in the wild.

Publications: Quarterly newsletter; biennial proceedings; *Trumpetings,* bimonthly.

Whitetails Unlimited (WTU)
P.O. Box 422
Sturgeon Bay, WI 54235
Phone: (414) 743–6777
Fax: (414) 743–4658

WTU was founded in 1982 to promote sound white-tailed deer management.

Publications: Chapter Connections, periodic newsletter; *The Deer Trail,* magazine, five issues per year; *Trail Talk,* annual; *WTU Insider,* periodic newsletter.

Whooping Crane Conservation Association (WCCA)
1007 Carmel Avenue
Lafayette, LA 70501
Phone (318) 234–6339

The WCCA was founded in 1961 to prevent the extinction of the whooping crane by managing a captive propagation program as a backup for the wild population and to support the establishment of a second wild population to reinforce survival against disaster.

Publications: Grus Americana, quarterly; annual membership directory.

Print Resources 7

Many books as well as scientific and popular articles address aspects of ecology, conservation, biodiversity, and endangered species. This chapter provides an annotated list of selected books and other publications that allow the reader to learn about endangered species and related topics.

Books

General Biology

Curtis, Helena, and Sue N. Barnes. *Biology.* New York: Worth Publishers, 1989. ISBN 0 87901 394 9. 1192 pp.

An excellent introduction to the important concepts of biology, including how plants and animals are classified and the definition of a species using evolution as a major theme. The unity of life, energetics, genetics, diversity of life, biology of plants, biology of animals, evolution, and ecology are all covered.

Endangered Species

Ackerman, Diane. *The Rarest of the Rare: Vanishing Animals, Timeless Worlds.* New York: Random House, 1995. ISBN 0 679 40346 9. 185 pp.

Ackerman argues that nature can and will take care of itself and that it is people we need to worry about because every extinction we cause brings us closer to the brink. Ackerman provides insight into her personal odyssey into the worlds of the monk seal, the short-tailed albatross, the golden lion tamarin, and the monarch butterfly, creatures that may well disappear before most of us can see them.

Beans, Bruce E. *Eagle's Plume: The Struggle to Preserve the Life and Haunts of America's Bald Eagle.* New York: Scribner's, 1996. ISBN 0 684 80696 7. 318 pp.

The bald eagle *(Haliaeetus leucocephalus)* is the symbol of our nation. But through hunting it and spraying crops and ornamental plants with DDT, we almost forced it into extinction. We failed to realize until it was almost too late that DDT builds up exponentially in the food chain. When an eagle eats prey that has been exposed to DDT, the eagle's body absorbs some of the DDT. One of the key effects of DDT is that it causes the shell of the eagle's egg to decrease in thickness, leading to breakage of eggs before the chicks inside can hatch. This book describes attempts to save the bald eagle.

Bergman, Charles. *Wild Echoes: Encounters with the Most Endangered Animals in North America.* New York: McGraw-Hill, 1990. ISBN 0 07 004922 X. 322 pp.

Bergman argues that although extinction is a natural phenomenon, extinctions that occur because of human activities are not natural. The author presents the stories of selected species, such as the gray wolf, the Florida panther, and the West Indian manatee. An appendix contains a partial list of the species that have become extinct since 1740.

Cone, Joseph. *A Common Fate: Endangered Salmon and the People of the Pacific Northwest.* New York: Henry Holt, 1995. ISBN 0 8050 2388 7. 340 pp.

The Pacific salmon has been around for 50 million years and in a handful of decades humans have driven it to the brink of extinction. This book highlights how the need for energy in the Pacific Northwest drove people to dam the very rivers that the salmon uses to live and especially to breed. It also details how overexploitation continues to threaten the existence of the salmon. The catch of salmon began to decline in 1894 and continued to decline through 1926. The Bonneville Dam was constructed in 1938 and the Grand Coulee Dam was completed in 1941. In 1969 the *U.S. v. Oregon* and *U.S. v. Washington* court cases determined that the Columbia Basin treaty tribes have a right to a fair and equitable share of the salmon harvest; that is, half of all the harvestable salmon destined for the tribes' traditional fishing places. The National Marine Fisheries Service began a formal review of the status of the salmon on the Columbia and Snake rivers under the ESA. The Northwest Electric Power Planning and Conservation Act of 1980 is to "protect, mitigate, and enhance" salmon runs affected by the hydroelectric development of the Columbia River. The Snake River Coho salmon became extinct in 1986. The Shoshone-Bannock Tribes of Idaho petitioned to protect the sockeye salmon of the Snake River in April 1990 under the ESA. The Snake River sockeye were listed as endangered in December 1991 and the Snake River chinook stocks were proposed to be listed.

Dary, David A. *The Buffalo Book.* New York: Avon Books, 1974. ISBN 0 380 00475 5, 374 pp.

No one knows precisely how many American buffalo *(Bison bison)* lived on the Great Plains when the Europeans arrived in North America. Ernest Thompson Seton, a well-known naturalist, set the number at between 30 and 75 million. A killing spree began in the 1830s and reached a peak in the 1870s. The animals were killed initially for their meat and skins, then for their skins and tongues, and ultimately only for their tongues, which were considered a delicacy. The rest of each carcass was left to rot on the plains. Dary provides an interesting overview of the buffalo, the buffalo trade, and efforts to save the buffalo.

DiSilvestro, Roger L. *The Endangered Kingdom: The Struggle to Save America's Wildlife.* New York: John Wiley and Sons, 1989. ISBN 0 471 60600 6. 241 pp.

DiSilvestro claims that with the exception of the house sparrow, the starling, and the brown rat, which thrive on contact with humans, most nonhuman organisms are threatened by human activities, especially because of habitat alteration and destruction.

This book provides a history of the relationship between hunters and wildlife managers and focuses on the effects on specific species of attempts to manage the species.

DiSilvestro, Roger L. *Fight for Survival.* New York: John Wiley and Sons, 1990. ISBN 0 471 50835 7. 284 pp.

Because of technology and our precipitously increasing population, according to DiSilvestro, we humans have become a force that is capable of changing the face of the globe. The author talks about global warming, acid rain, and the thinning of the ozone layer and their impact. He also discusses what ordinary people can do to help save our globe and its inhabitants, both human and nonhuman.

Durrell, Lee. *State of the Ark.* Garden City, NY: Doubleday, 1986. ISBN 0 385 23668 9. 224 pp.

In this heavily illustrated book, Durrell describes the biosphere ("the ark") and the complex interactions of its parts. He also describes environmental problems encountered in each of the world's regions and the state of the world's species and their interactions.

Fitzgerald, Sarah. *Whose Business Is It?* Washington, DC: World Wildlife Fund, 1989. ISBN 0 942635 13 2. 457 pp.

The World Wildlife Fund was founded to help protect wildlife, to monitor worldwide trade in wildlife products, and to educate trade professionals and the general public about the trade in wildlife and its impacts. In her book, Fitzgerald provides an overview of this trade, including which species are involved and their status. She also provides a detailed analysis of the Convention on International Trade in Endangered Species of Wild Fauna and Flora, commonly called CITES.

Gipps, J. H. W., ed. *Beyond Captive Breeding.* New York: Oxford University Press, 1991. ISBN 0 19 854019 1. 284 pp.

This book is the proceedings of a meeting of scientists who capture wild mammals that are threatened with extinction, bring them to zoos and other locations, and attempt to breed them in captivity. The ultimate goal is to reintroduce these animals into their natural habitats when conditions are more favorable for their survival.

Groombridge, Brian. *1996 IUCN Red List of Threatened Animals.* Cambridge, UK: World Conservation Union, 1997. ISBN 2 8317 0335 2. 308 pp.

The "Red Book" is a list of threatened and extinct species throughout the world; it also includes threatened genera and subspecies. Species that are threatened because of commercial interest (trade) are also listed.

Mann, Charles C., and Mark L. Plummer. *Noah's Choice: The Future of Endangered Species.* New York: Alfred A. Knopf, 1995. ISBN 0 679 42002 9. 303 pp.

Mann and Plummer suggest that the current Endangered Species Act leads to two stages of action. First, scientists examine a species to determine if it should be listed. Second, if the species is listed, government agencies design and enact a program to conserve it. The authors believe that the process should be modified to separate the two stages. The listing process would be essentially an informational device. It would be used to take an aggressive inventory of the nation's national heritage at the regional and ecosystem level. Mann and Plummer suggest that "take" be redefined as intentional, direct harm to an individual member of a list species. They believe we can provide greater protection to larger numbers of species in this manner.

Nilsson, Greta. *The Endangered Species Handbook.* Washington, DC: Animal Welfare Institute, 1993. LCN 82 072956. 245 pp.

Nilsson provides a discussion of how and why species are becoming endangered and ultimately extinct. She focuses on the role humans play, especially the trade in animals for their fur and skin and as pets, as well as trophy and meat hunting. While direct human acts, such as hunting, cause problems for many species, other species fall victim to human acts not intended to cause harm, such as using pesticides, producing pollution that leads to acid rain, and disposing of industrial refuse that contains PCBs. Nilsson also describes legislation at the state, federal, and international levels. The text of the Convention on International Trade in Endangered Species of Wild Fauna and Flora is presented in an appendix, as are lists of extinct animals and animals that are threatened with extinction. The author also lists resources that can be used by citizens, teachers, and anyone else in a position to teach others about the plight of endangered species of animals.

Schorger, A. W. *The Passenger Pigeon.* Madison: University of Wisconsin Press, 1955. LCN 54 6738. 424 pp.

Before the 1880s, flocks of passenger pigeons were immense. It is estimated that the Pennsylvania flock contained 200 million individuals. The last specimen was taken from the wild in Sargents Pick County, Ohio, on March 24, 1900. The pigeon was used for food, with squabs (young pigeons) considered a delicacy. It was also used in trap shooting.

Whales in Captivity: Right or Wrong? Proceedings of a Symposium. Ottawa, Ontario: Canadian Federation of Humane Societies, 1990.

Clearly, we have the power to kill or capture and hold whales. But do we, as a species, have a "right" to do this to another intelligent species?

World Wildlife Fund. *The Official World Wildlife Fund Guide to Endangered Species of North America. Volume 1: Plants and Mammals.* Washington, DC: Beacham, 1990. ISBN 0 933833 17 2. 1536 pp.

This book provides a species-by-species description of the plants and animals listed as threatened or endangered under the Endangered Species Act. Each description of an animal contains a photograph; the scientific and common name(s) of the animal; a physical description; descriptions of the animal's behavior, habitat, historic range, and current distribution; the animal's conservation and recovery plan; a bibliography; and a thumbnail summary of the animal. The book also lists a government point of contact for each species, usually at the field office level of the U.S. Fish and Wildlife Service. The volume contains several appendixes, including an evaluation of listed species as improving, stable, declining, extinct, or unknown. Another appendix presents species that are likely to be listed in the near future.

Historical Geology and the History of the Earth

Dunbar, Carl O. *Historical Geology.* New York: John Wiley and Sons, 1949. 573 pp.

This textbook provides an excellent and easy-to-understand introduction to the geological ages. It is well illustrated and provides an overview of the fossil record and how it is used to understand the progress of life on Earth. The book describes the species that were dominant during each age.

Hartmanh, William K., and Ron Miller. *The History of Earth: An Illustrated Chronicle of an Evolving Planet.* New York: Workman, 1991. ISBN 1 56303 122 2. 260 pp.

This book provides a well-illustrated overview of the geological ages and what occurred during each one. Hartmanh and Miller describe Earth before life appeared and when life emerged, the evolution of living organisms, and the transitions from one dominant group of species to the next.

Extinction

Balouet, Jean-Christophe, and Eric Alibert. *Extinct Species of the World.* New York: Barron's, 1990. ISBN 0 8120 5799 6. 192 pp.

Living organisms are commonly grouped hierarchically into kingdom, class, order, family, genus, and species, and this classification scheme attempts to reflect species' origins and evolution. Based on the fossil record, species typically exist for 1 to 10 million years, although some exist much longer, and some become extinct much sooner. Balouet and Alibert provide a historical review of efforts to preserve species. The authors point out that species that have gone extinct are often ones that humans derive a profit from or species that have suffered from competition or predation of introduced species. The book provides an overview of extinct species around the world.

Cadieux, Charles L. *Extinction.* Washington, DC: Stone Wall Press, 1991. ISBN 0 913276 59 6. 259 pp.

Cadieux provides text and drawings that highlight specific animals and groups of animals worldwide that are threatened with extinction; the author also explains what people are doing to try to save these animals. For example, the number of Attwater's prairie chickens *(Tympanuchus cupido attwaterii)* has been decreasing steadily since the 1930s. The animal's ancestral habitat is threatened by the expansion of Houston. In 1980, the Attwater's Prairie Chicken National Refuge was established to protect the habitat of approximately 150 chickens, and a portion of the Arkansas National Wildlife Refuge was set aside to protect another population of these birds. The number of these chickens, however, has continued to drop, and the species is well on the road to extinction, according to Cadieux. The author supplies an appendix listing people who seek to protect endangered species.

Catton, William R., Jr. *Overshoot.* Urbana: University of Illinois Press, 1980. ISBN 0 252 00818 9. 298 pp.

Catton claims that we humans, the most powerful species in the world, have increased in numbers beyond the carrying capacity of our finite habitat. This book advocates a paradigm shift, a change in the kinds of questions we ask about the environment and our impact on it. The author also suggests how to begin this paradigm shift.

Erickson, Jon. *Dying Planet: The Extinction of Species.* Blue Ridge Summit, PA: Tab Books, 1991. ISBN 0 8306 6726 1. 188 pp.

This book provides an introduction to the origin and history of life on Earth. The book describes the mass extinctions that have occurred and what scientists believe caused these extinctions, including asteroid collisions, volcanic eruptions, ice ages, greenhouse warming, and human activity.

Erwin, Douglas H. *The Great Paleozoic Crisis: Life and Death in the Permian.* New York: Columbia University Press, 1993. ISBN 0 231 074662 2. 327 pp.

The Permian Age was named for the town of Perm in the Ural Mountains of Russia. By the end of the Permian period, it is estimated, 57 percent of all families had perished, and 83 percent of genetic diversity was gone. Erwin describes the possible scenarios that might have led to this mass extinction.

Fitter, Richard, and Maisie Fitter, eds. *The Road to Extinction.* Gland, Switzerland: IUCN Publications, 1987. ISBN 2 88032 929 9.

The proceedings of a symposium hosted by the International Union for Conservation of Nature and Natural Resources (IUCN), this book describes the history and uses of the IUCN's Red Data Books, a register of threatened wildlife. The publication also defines the 13 milestones on the road to extinction.

Glen, William, ed. *The Mass-Extinction Debates: How Science Works in a Crisis.* Stanford, CA: Stanford University Press, 1994. ISBN 0 8047 228 62. 370 pp.

The history of life on Earth has not been smooth, but rather has been punctuated by a series of mass extinctions in which many species disappeared in a relatively short time. The most famous is probably the K-T mass extinction, during which the dinosaurs went extinct. Researchers have put forth a number of theories to account for these mass extinctions, with two predominating. The first involves the impact of an extraterrestrial object on Earth, and the second, an increase in volcanic activity. This book provides a forum for debate among the key theorists in each camp.

Kaufman, Les, and Kenneth Mallory, eds. *The Last Extinction.* Cambridge, MA: MIT Press, 1986. ISBN 0 262 11115 2. 208 pp.

The contributors provide a discussion of how and why extinctions seem to be on the increase, primarily as a result of human activity.

Raup, David M. *Extinction: Bad Genes or Bad Luck?* New York: W. W. Norton, 1991. ISBN 0 3933 0927A. 210 pp.

Individuals are born, live for a variable amount of time, and then die. The same is true for species. They arise, they successfully fill ecological niches, and ultimately most disappear. Raup and most other experts who have studied the problem suggest that 99.9 percent of all species that ever existed have become extinct. Consider, for example, the trilobites, crablike organisms that dominated the ocean bottoms for more than 350 million years before disappearing in a mass extinction 245 million years ago. They do not appear to have left any descendants. Were they victims of bad genes or bad luck? Raup discusses the probable causes of extinction and mass extinctions. He discusses the K-T (Cretaceous-Tertiary) mass extinction, during which virtually all plant and animal groups lost some species and genera. On land, dinosaurs were the hardest hit. Raup provides models that can help readers understand extinction and its potential biological and physical causes.

Regenstein, Lewis. *The Politics of Extinction: The Shocking Story of the World's Endangered Species.* New York: Macmillan, 1975. ISBN n/a. 280 pp.

Regenstein discusses the politics involved in dealing with such diverse species as the wolf, the grizzly bear, the prairie dog, the wild horse, and the cougar. Who speaks for these animals, and who speaks against them?

Ward, Peter Douglas. *On Methuselah's Trail: Living Fossils and the Great Extinctions.* New York: W. H. Freeman, 1992. ISBN 0 7167 2203 8. 212 pp.

Living fossils are small groups of animals or plants that are the only living representatives of geologically ancient life-forms and that have not changed in the tens to hundreds of millions of years since the majority of their fellow life-forms disappeared. The horseshoe crab (*Limulus* spp.) is a classic example. Although it looks crablike, the horseshoe crab is more closely related to scorpions and spiders. Over 300 million years ago, there were hundreds of species of horseshoe crabs and their relatives, the sea scorpions. Today, sea scorpions are extinct, and only four species of horseshoe crabs remain.

The Paleozoic, Mesozoic, and Cenozoic eras were defined by geologist John Phillips and correspond to the great divisions of life on Earth. The plants and animals that are characteristic of one era are unlike those that characterize another. A mass extinction is defined as a period of less than 1 million to 15 million years during which a large number of species and higher taxa become extinct. Most experts acknowledge five mass extinctions. The Late Ordovician, when about 22 percent of the existing families disappeared; the Devonian, when a similar percentage of families disappeared; the Permian, when 50 percent of the existing families (and 76–96 percent of all species) disappeared; the Late Triassic, when 20 percent disappeared, and the Late Cretaceous, when 15 percent disappeared. Note that all of these mass extinctions occurred millions of years before humans appeared.

Swanson, Timothy M. *The International Regulation of Extinction.* New York: New York University Press, 1994. ISBN 0 8147 7992 1. 289 pp.

Can humans limit their population growth and the spread of their influence? Swanson argues that if we do not limit the spread of our influence, we will have a significant and negative impact on biodiversity.

Tudge, Colin. *Last Animals at the Zoo: How Mass Extinction Can Be Stopped.* Washington, DC: Island Press, 1992. ISBN 1 55963 158 9. 263 pp.

Tudge argues that zoos can provide support for endangered species by breeding animals in captivity to prevent them from going extinct and to provide populations of animals that might be re-released into their natural habitat. It is likely that zoo breeding is the only way to prevent some species, such as the rhino and the tiger, from becoming extinct. Tudge points out that this is not as simple as putting a male and a female in a cage together and waiting for the inevitable. Many species require specific environmental conditions to breed—for example, a specific number of animals present or a specific mix of males and females present. The author also cautions that breeding must be managed to avoid inbreeding and other genetic disasters.

Ward, Peter D. *The Call of Distant Mammals: Why the Ice Age Mammals Disappeared.* New York: Copernicus Books, 1997. ISBN 0 387 94915 1. 241 pp.

The large Ice Age mammals, such as the woolly mammoth, the mastodon, the woolly rhinoceros, the giant sloth, and the sabertoothed cats, all disappeared relatively recently in the geologic past and in a relatively short time period. Ward maintains that a great catastrophe occurred for those mammals—the appearance of humans.

Biodiversity

Dixon, John A., and Paul B. Sherman. *Economics of Protected Areas: A New Look at Benefits and Costs.* Washington, DC: Island Press, 1990. ISBN 1 55693 032 9, 235 pp.

Natural areas that are relatively undisturbed by people are an increasingly scarce resource in both developed and developing countries. This book describes the economic issues associated with the establishment and management of protected areas. A second section provides examples of how approaches developed in the first section are used, including an analysis of specific benefits.

Hoose, Phillip M. *Building an Ark: Tools for the Preservation of Natural Diversity through Land Protection.* Covelo, CA: Island Press, 1981. ISBN 0 933280 09 2. 221 pp.

Hoose argues that biodiversity is a vital concern to humans. If, either through action or inaction, we allow too many species to perish, we will certainly follow, he believes. For example, 90 percent of the world's diet comes from just 20 plants, with 3 being the mainstays: wheat, corn, and barley. These crops are generally grown in monoculture. Eventually, diseases and pests adapt to each modern hybrid, and new plant hybrids must be created. If wild progenitors are not available, plant breeders cannot breed new hybrids—a potential disaster.

Pearce, David, and Dominic Moran. *The Economic Value of Biodiversity.* London: Earthscan Publications, 1994. ISBN 1 85383 195 6. 172 pp.

Economic forces have, directly and indirectly, been the driving force behind the extinction of many of the world's biological resources. But genetic and species diversity are important for the long-term survival of humans and other organisms. Pearce and Moran provide an economic analysis of the value of biodiversity, including direct and indirect valuation approaches, economic estimates, and an examination of the economic effects of development alternatives.

Weiner, Jonathan. *The Beak of the Finch.* New York: Alfred A. Knopf, 1994. ISBN 0 679 40003 6. 332 pp.

This book describes the work of Peter and Rosemary Grant and their associates, who study the finches of Daphne Major, one of the Galápagos Islands. Charles Darwin made the finches and the Galápagos Islands famous in his classic books on the theory of evolution. Weiner's book describes the Grants' discoveries about the rapid changes that occur in these finches in response to changes in their physical environment, especially their food supply. The book also provides an excellent bibliography.

Wilson, Edward O. *The Diversity of Life.* New York: W. W. Norton, 1992. ISBN 0 393 31047 7. 424 pp.

Wilson, a professor at Harvard University and a leading proponent of sociobiology, provides a readable discussion of the forces of evolution and the causes of adaptive radiation, which lead to new species. Wilson posits that much of the biosphere remains to be explored and that we do not even know the total number of species that exist on Earth, but he suggests that more than 1.4 million species currently exist, more than half of those arthropods. He provides a map and a discussion of the world's hot spots, places where many species exist that are found nowhere else and that are in danger of extinction because of human activity. Wilson argues eloquently that if we allow ourselves to damage or destroy these and other ecosystems, in the process decreasing biodiversity, the consequences could be catastrophic and irreversible.

Conservation

Allin, Craig W. *The Politics of Wilderness Preservation.* Westport, CT: Greenwood Press, 1982. ISBN 0 313 21458 1. 304 pp.

Prior to the turn of the century, many Americans believed that they could misuse and overuse their land, then abandon it and move on, usually farther west, into the wilderness. After the turn of the century, it became clearer and clearer that the wilderness was rapidly disappearing and that some effort must be made to preserve it. Allin describes the events that led to the Wilderness Protection Act and how it is implemented.

Burgman, M. A., S. Ferson, and H. R. Akcakaya. *Risk Assessment in Conservation Biology.* New York: Chapman and Hall, 1993. ISBN 0 412 35030 0. 314 pp.

All species, including humans, live in uncertain environments. Their long-term survival depends on short-term probabilistic fluctuations in populations. Risk—the topic of this book—is the probability of an adverse outcome. Risk assessment involves qualitative and quantitative methods for measuring risk. The book demonstrates how population growth can be defined by the risks associated with phenotypic, demographic, environmental, and spatial variation.

Ditton, Robert B., and Thomas L. Goodale, eds. *Environmental Impact Analysis: Philosophy and Methods.* Madison, WI: Sea Grant Publications, 1972. 171 pp.

One of the requirements of the National Environmental Protection Act is that the branches, bureaus, and agencies of the federal government must determine, via environmental impact statements, how their activities impact the environment. Ditton and Goodale provide an overview of the method and philosophy of doing environmental impact statements and their effect on protecting the environment.

Douglas, William O. *A Wilderness Bill of Rights.* Boston: Little, Brown, 1965. LCN 65 21350. 192 pp.

Justice Douglas cites the Wilderness Act, which states, "A wilderness, in contrast with those areas where man and his own works dominate the landscape, is hereby recognized as an area where the earth and its community of life are untrammeled by man, and where man himself is a visitor who does not remain." In a compelling book, Douglas maintains that a wilderness is more valuable than the price that can be placed on the cellulose in its trees, the board-feet of its timber, or the hydroelectric power of its rivers and, as such, must be preserved for its own value.

Environmental Department. *Environmental Assessment Sourcebook. Volume I: Policies, Procedures, and Cross-Sectoral Issues.* Washington, DC: World Bank, 1991. ISBN 0 8213 1843 8. 216 pp.

The World Bank (the International Bank for Reconstruction and Development) is a specialized United Nations agency established at the Bretton Woods Conference in 1944. Its mandate is "to assist in the reconstruction and development of territories of members by facilitating the investment of capital for productive purposes [and] to promote private foreign investment by means of guarantees or participation in loans [and] to supplement private investment by providing, on suitable conditions, finance for productive purposes out of its own capital." This book describes the environmental assessment procedures the bank requires before it considers a project for funding.

Erickson, Brad, ed. *Call to Action.* San Francisco, CA: Sierra Club Books, 1990. ISBN 0 87156 611 7. 250 pp.

This book provides an overview of the Sierra Club's views on animals, endangered species, and the methods that should be used to preserve and protect them.

Ives, Richard. *Of Tigers and Men: Entering the Age of Extinction.* New York: Doubleday, 1996. ISBN 0 385 47816 X. 304 pp.

Ives presents a vivid description of the efforts to save the wild tigers of India, Nepal, and Southeast Asia.

Olney, P. P. J. S., G. M. Mace, and A. T. C. Feistner, eds. *Creative Conservation: Interactive Management of Wild and Captive Animals.* New York: Chapman & Hall, 1994. ISBN 0 412 49570 8. 517 pp.

This symposium describes methods currently used to prevent

species of animals from becoming extinct. Techniques include the relocation of species to new habitats, captive-breeding programs, and the reintroduction of captive-bred animals into the wild. The authors provide examples of successful programs.

Reiger, John F. *American Sportsmen and the Origins of Conservation.* Norman: University of Oklahoma Press, 1986. ISBN 0 8061 2021 5. 316 pp.

The Boone and Crockett Club was formed by Theodore Roosevelt, George Bird Grinnell, and other prominent sportsmen in 1887. The club was the first private organization to deal with conservation issues on a national level. This book provides a detailed history of the relationship between sports hunters and fishermen and conservation.

Remmert, Hermann, ed. *Minimum Animal Populations.* New York: Springer-Verlag, 1994. ISBN 3 540 56684 8 and 0 387 56684 8. 156 pp.

One of the most compelling questions in dealing with species that are thought to be threatened or endangered is how small can a population be and still be viable? That is, what is the minimum number of breeding individuals that can maintain a species without running into problems of inbreeding?

Ronald, Ann, ed. *Words for the Wild.* San Francisco: Sierra Club Books, 1987. ISBN 0 87156 709 1. 365 pp.

This book consists of a series of reading by Emerson, Thoreau, and others that the Sierra Club views as champions of the environment and endangered species.

Sharpe, Grant W. *Interpreting the Environment.* New York: John Wiley, 1982. ISBN 0 471 09007 7. 694 pp.

Sharpe contends that if we want to save the environment, prevent habitats from being destroyed, and protect plants and animals from becoming threatened or endangered, then we must involve nonspecialists. The author presents the methods used to explain the problems so that anyone can understand them

Stroud, Richard H., ed. *National Leaders of American Conservation.* Washington, DC: Smithsonian Institution Press, 1985. ISBN 0 87474 867 4. 432 pp.

This book provides biographies of the leaders of the American conservation movement, arranged in alphabetical order.

Trefethen, James B. *An American Crusade for Wildlife.* New York: Winchester Press, 1975. ISBN 0 87691 207 2. 409 pp.

Trefethen explores the early exploitation of America's wildlife by market hunters who wiped out the passenger pigeon and almost wiped out the buffalo. He describes how the tide changed with the development of wildlife management.

Turner, Tom. *100 Years of Protecting Nature.* San Francisco, CA: Harry N. Abrams, Sierra Club. 1991. ISBN 0 8109 3820 0. 288 pp.

The Sierra Club, started by John Muir and others in 1892, has been active in the environmental movement, shaping public opinion, lobbying, and especially litigating. Turner also provides a chronology of the activities of the club. For a contrasting view on the Sierra Club, see Ron Arnold's book, listed below in the Ecology section of this chapter.

———. *Wild by Law.* Sierra Club Legal Defense Fund, 1990. ISBN 0 87156 627 3. 154 pp.

This book provides the Sierra Club's view of its work on the environment, conservation, and the protection of endangered species.

Usher, Michael B., ed. *Wildlife Conservation Evaluation.* New York: Chapman and Hall, 1986. ISBN 0 412 26750 0.

Evaluations of wildlife are performed to gain knowledge about a specific species or groups of species and to determine if a particular piece of land can be used by that species.

Ecology

Arnold, Ron. *Ecology Wars: Environmentalism As If People Mattered.* Bellevue, WA: Free Enterprise Press, 1987. ISBN 0 939571 00 5. 182 pp.

Arnold is clearly not a fan of the environmental movement's group of ten, the oldest and largest environmental organizations in the United States. They include the National Wildlife Foundation, the National Audubon Society, the Wilderness Society, the

Natural Resources Defense Council, the Environmental Defense
Fund, the Izaak Walton League of America, the National Parks
and Conservation Association, Friends of the Earth, and the Environmental Policy Institute. Arnold calls these organizations
"volunteer struggle groups." He credits John Muir, the founder
of the Sierra Club, with originating the idea. Arnold points out
that these organizations function in three areas: the shaping of
public opinion, legislative lobbying, and litigation. Political
awareness is the hallmark of the organizations. For a contrasting
view on the Sierra Club, see the entries for Tom Turner's books in
the Conservation section of this chapter.

Begon, Michael, and Martin Mortimer. *Population Ecology.*
Boston: Blackwell Scientific Publications, 1986. ISBN 0 632 01443
1. 220 pp.

This book provides an analysis of the key concepts of population
ecology, including what is a population, intra- and interspecies
competition, predation, population regulation, and community
structure. Many of these concepts are central to the decision of
whether a species is or is not in danger of becoming extinct.

Carson, Patrick, and Julia Moulden. *Green Is Gold: Business
Talking to Business about the Environmental Revolution.* New
York: HarperBusiness, 1991. ISBN 0 88730 520 2. 216 pp.

Businesspeople are discovering that increasing numbers of consumers are concerned about the environment and are using their
pocketbooks to demonstrate that concern. This book describes
how a business can develop and implement a "green" strategy. It
provides a list of books and environmental experts that can help
a business develop green plans and products.

Coleman, Daniel A. *Ecopolitics.* New Brunswick, NJ: Rutgers
University Press, 1994. ISBN 0 8135 2054 1. 236 pp.

Coleman suggests that the environmental movement has many
starting points: the publication of *Walden* by Thoreau in 1854; the
formation of the Sierra Club in 1892; the publication of *Silent
Spring* by Rachel Carson in 1962; and the first Earth Day in 1970,
which led to the Clean Air Act, the Clean Water Act, and the
founding of the Environmental Protection Agency. Whichever
you choose, Coleman provides an insightful introduction to the
environmental movement and green social and political issues.

Elkington, John, Julia Hailes, and Joel Makower. *The Green Consumer.* New York: Penguin Books, 1988. ISBN 0 1401 2708 9. 342 pp.

First and foremost, a green product is not dangerous to the health of humans or animals. A green product is one that does not damage the environment or use disproportionate amounts of energy during manufacture, use, or disposal. It does not cause unnecessary waste because of excessive packaging or a short life. A green product does not cause unnecessary use of or cruelty to animals and does not use materials derived from threatened or endangered species or environments. This book provides an overview of environmental problems and how human activities affect them. Individual chapters describe how to shop for environmentally friendly products, including automobiles, groceries, garden and pet supplies, home energy products, furnishings, cosmetics and personal care products, and travel. The authors also provide an overview of how to get involved, a list of environmental books, and an annotated list of environmental organizations, as well as a list of state and federal environmental offices.

Emmel, Thomas C. *Population Biology.* New York: Harper and Row, 1976. ISBN 0 06 041904 0. 371 pp.

An ecosystem includes the physical environment and all its component plant and animal populations. Emmel describes the processes that control the ebb and flow of populations of organisms and their interrelationships.

Environmental Action. *Earth Day—The Beginning.* New York: Arno Press and the New York Times, 1970. ISBN 0 405 02712 5. 277 pp.

Earth Day, which began as a student-led campus movement, was initially observed on 21 March 1970 and was first observed internationally on 22 April 1970. The purpose of Earth Day was to focus public attention on the necessity of the conservation of the world's natural resources. Earth Day provides a forum for summing up current environmental problems, such as the pollution of air, water, and soil; the depletion of nonrenewable resources; and the destruction of habitat, which results in the loss of species of plants and animals. Earth Day emphasizes solutions that slow and possibly reverse the negative effects of human activities. This book provides short essays about these issues.

Hinckley, Alden Dexter. *Ecology.* New York: Macmillan, 1976. ISBN 0 02 354551 8. 342 pp.

Hinckley provides an overview of ecology, the study of relationships between organisms and between organisms and their environment. The author also provides a review of human impact on the environment.

Isard, Walter. *Ecologic-Economic Analysis for Regional Development.* New York: Free Press, 1972. ISBN n/a. 270 pp.

This book discusses the economic basis for regional development and provides a case study of a developmental project.

Killingsworth, M. Jimmie, and Jacqueline S. Palmer. *Ecospeak.* Carbondale: Southern Illinois University Press, 1992. ISBN 0 8093 1750 8. 312 pp.

Killingsworth and Palmer provide an interesting discussion of how we talk and write about the environment, environmental problems, and ecology.

Knudsen, Jens W. *Biological Techniques.* New York: Harper & Row, 1966. LCN 66 10839. 525 pp.

This book describes the methods that scientists use to collect and preserve plants and animals so they can be studied.

Krebs, J. R., and N. B. Davies, eds. *Behavioral Ecology.* Sunderland, MA: Sinauer Associates, 1978. ISBN 0 87893 433 2. 494 pp.

This relatively technical book provides an overview of ecology and the effects of human and animal behavior on those organisms' interactions with the environment.

Lauenroth, William K., Gaylord V. Skogerboe, and Marshall Flug, eds. *Analysis of Ecological Systems: State-of-the-Art in Ecological Modeling.* New York: Elsevier, 1983. ISBN 0 444 42179 3. 172 pp.

This book is the proceedings of a professional meeting on ecological modeling. Modeling is and will continue to be an essential component of the study of how and why species become endangered. Modeling and understanding models require a strong background in science and mathematics.

Leopold, Aldo. *A Sand County Almanac.* New York: Ballantine Books, 1966. ISBN 0 345 34505 3. 295 pp.

Aldo Leopold (1887–1948) is considered the father of wildlife ecology and was a renowned scientist and scholar, an exceptional teacher, a philosopher, and a gifted writer. His book *A Sand County Almanac* has been heralded as the century's literary landmark in the field of conservation. His keen observations of the natural world are described in exceptional, poetic prose, and the book's philosophy has guided many to discover what it means to live in harmony with the land and with one another.

Leopold attended Yale, then joined the U.S. Forest Service and was assigned to the Arizona Territories. While in Arizona, he began to see the land as a living organism and to develop the concept of community. Leopold's cornerstone book, *Game Management* (1933), defined fundamental skills and techniques for managing and restoring wildlife populations and created a new science combining forestry, agriculture, biology, zoology, ecology, education, and communication. He developed the ideas in this book while working an abused and abandoned farm he purchased in central Wisconsin during the Depression. The book was published posthumously in 1949. In 1948, before his work achieved public acceptance, Leopold died helping a neighbor fight a brush fire.

Mitchell, John G., and Constance L. Stallings. *Ecotactics: The Sierra Club Handbook for Environmental Activists.* New York: Pocket Books, 1970. ISBN 067 127 0664. 288 pp.

Mitchell and Stallings define *ecotactics* as the "science of arranging and maneuvering all available forces in action against enemies of the earth." They promote activities that they believe will help save Earth and provide lists of organizations and professional societies that will help in this effort.

National Research Council. *Ecological Knowledge and Environmental Problem-Solving: Concepts and Case Studies.* Washington, DC: National Academy Press. ISBN 0 309 03645 3. 388 pp.

This book discusses ecological concepts, then applies them to a series of case studies. Each case study includes an introduction to the issues, an overview of the specific problem, and the approach used to deal with it. Each also illustrates how ecological knowledge is used to develop management guidelines and how planning the guidelines unfolds. The publication also includes a bibliography.

Odum, Eugene P. *Fundamentals of Ecology.* Philadelphia: W. B. Saunders, 1971. ISBN 0 7216 6941 7. 574 pp.

In this book, Odum provides an introduction to the concepts of ecology, focusing on ecosystems and the energy flow through them. He also develops the concept of species and the role of the individual and the species in the ecosystem. The book provides insights into freshwater, marine, estuarine, terrestrial, microbial, and human ecology.

Oelschlaeger, Max, ed. *After Earth Day: Continuing the Conservation Effort.* Denton: University of North Texas Press, 1992. ISBN 0 929398 40 8. 241 pp.

This book contains a series of essays on the theory and practice of conservationism and environmentalism.

Owen, Oliver S. *Natural Resource Conservation: An Ecological Approach.* New York: Macmillan, 1975. ISBN 0 02 390020 2. 883 pp.

Owen provides an overview of ecological concepts and their relationship to conservation of natural resources. He describes the complex interrelationships between soil, water, energy generation and use, and pollution. Owen argues that we have to accept that the number of "wide-open spaces" is rapidly diminishing, so a farmer can no longer work land to exhaustion and then move on, and factories can no longer pollute the environment and then move on. He also argues that we must accept the concept of a "spaceship earth" that has limited and mostly nonrenewable resources, and that if we create pollution, we have to live with its effects.

Patten, Bernard C., ed. *Systems Analysis and Simulation in Ecology.* New York: Academic Press, 1976. ISBN 0 12 547204 8. 593 pp.

This is a difficult book that describes how scientists model ecosystems, including human interactions with the environment.

Ricklefs, Robert E. *Ecology.* New York: Chiron Press, 1979. ISBN 0 913462 07 1. 966 pp.

Ecology is the study of organisms, including humans, and their interactions with their environments. Ricklefs provides a survey of biological communities and community ecology as well as the process of phyletic evolution and speciation. The concept of reproductive isolation is also described.

Rifkin, Jeremy, ed. *The Green Lifestyle Handbook.* New York: Henry Holt, 1990. ISBN 0 8050 1369 5. 198 pp.

Rifkin provides an overview of the environmental problems that concern Greens and the solutions that Greens believe will work. Other authors address home energy use and cleaning products, grocery shopping, how to live the green lifestyle, and how to get organized. Appendixes describe the environmental crisis, solutions that the Greens believe are unsound, such as nuclear power and genetic engineering, and an annotated guide to books on ecology.

Risser, Paul G., and Kathy D. Cornellson. *Man and the Biosphere.* Norman: University of Oklahoma Press, 1979. ISBN 0 8061 1610 2. 109 pp.

This book describes the Man in the Biosphere Program, an intergovernmental program that attempts to focus research, public education, and technical training on the biosphere and our complex interactions with it. The authors also describe biosphere reserves that currently exist in the United States. Each description includes the geographical location of the bioreserve, its altitude, its area, its physical features, its vegetation, its fauna, and the modifications made to it by humans. The book also lists who owns the land and estimates the scientific research potential of each reserve.

Shabecoff, Philip. *A Fierce Green Fire.* New York: Hill and Wang, 1993. 352 pp.

Ernst Haeckel, in 1866, coined the term *ecology* from the Greek *oikos* to describe the study of how organisms interact with each other and a shared environment. Aldo Leopold states that "all ethics rest upon a single premise: that the individual is a member of a community of interdependent parts." And this interdependence is the centerpiece of modern environmentalism. If we destroy the land, the air, the water, the plants, the animals, and even the microorganisms, we will soon follow suit. Shabecoff traces the beginnings of conservationism from the Forest Reserve Act of

1891 to Pinchot and Theodore Roosevelt and describes the swing of the environmental pendulum back and forth and back again.

Skalski, John R., and Douglas S. Robson. *Techniques for Wildlife Investigations: Design and Analysis of Capture Data.* San Diego: Academic Press, 1992. ISBN 0 12 647675 6. 237 pp.

One of the ways scientists learn about animal populations is the process of capture-mark-recapture. That is, an organism is captured in a live trap, carefully examined, marked in some way (for example, a bird typically has a metal or plastic band placed around its leg, and a mammal might have a collar placed around its neck), then released. Some time later, animals are recaptured in the live traps. The relative number of marked versus non-marked animals provides an indication of the size of the local population of the animal. This book provides a detailed description of how field trials should be designed and the statistical methods used to evaluate these trials.

Smith, Robert Leo. *Ecology and Field Biology.* New York: Harper and Row, 1966. LCN 66 15675. 686 pp.

Smith provides a good introduction to ecology and the concepts of ecosystem and community, as well as a good description of aquatic and terrestrial habitats. His discussion of species and how they arise is excellent.

————. *Ecology and Field Biology.* 2d ed. New York: Harper and Row Publishers, 1974. ISBN 0 06 046334 1. 850 pp.

An examination of the two editions of this book (see the preceding entry), which came out just eight years apart, shows how rapidly a scientific discipline changes its emphasis. In the second edition, Smith devotes a chapter to social behavior in populations and the importance of display behavior in controlling reproductive behavior. He also devotes a chapter to population genetics, natural selection, and speciation.

Radical Ecology

Foreman, Dave. *Confessions of an Eco-Warrior.* New York: Harmony Books, 1991. ISBN 0 517 58123 X. 229 pp.

Foreman is one of the cofounders of Earth First!, an environmental organization dedicated to the preservation and restoration of wilderness areas. He is a real-life proponent of "monkey wrenching," including tree spiking. The latter consists of inserting a large nail or other piece of metal or a hard nonmetallic object into a tree so that the trees will damage saws if they are put through sawmills. He claims that timber companies, the U.S. Forest Service, and other agencies are warned.

Scarce, Rik. *Eco-Warriors.* Chicago: Noble Press, 1990. ISBN 0 9622683 3 X. 291 pp.

The new environmentalists, sometimes called radical environmentalists, differ from their predecessors in that radical environmentalists abhor compromise. They are activists; they sit down in front of a bulldozer or climb up in a tree to prevent it from being cut down. They claim that their cause is not noble, but essential. They do not rule out monkey wrenching. Greenpeace, Earth First!, and the Sea Shepherds are prominent radical environmentalist groups.

Domestication

Clutton-Brock, Juliet. *A Natural History of Domesticated Mammals.* Austin: University of Texas Press, 1989. LCN 80 71080.

Clutton-Brock quotes Sir Francis Galton's essay about human domination and manipulation of the animal kingdom and restates the six conditions Galton felt were required to allow an animal to become domesticated: It should be hardy, have an inborn liking for man, be comfort loving, be useful to "savages," breed freely, and be easy to tend.

Clutton-Brock divides domestic animals into four large categories: (1) man-made animals, which include dogs, sheep, goats, cattle, and pigs as well as horses, asses, and mules; (2) exploited captives, which include cats, elephants, camels and llamas, reindeer, and Asiatic cattle; (3) small mammals, which include rabbits and ferrets as well as rodents and carnivores that are exploited for their fur; and (4) animals that are used in game ranching, such as deer and bovids. The author also describes the process of domestication for each group and discusses the ungulates that were exploited by pre-Neolithic humans.

Clutton-Brock, Juliet, ed. *The Walking Larder: Patterns of Domestication, Pastoralism, and Predation.* Boston: Unwin Hyman, 1989. ISBN 0 04 445013 3. 368 pp.

The authors of this book describe the manifold relationships between humans and animals, both in the past and in the present, and how these relationships affect the process of domestication. Early humans were constantly on the move, trying to expand their territory and find new resources. These early humans, like modern humans, had to have a source of protein. They found it in the wild animals that they followed and hunted, as well as the domestic animals that they used as a store of meat on the hoof—that is, as livestock. The book describes the development of pastoralism in Europe, Asia, and Africa. It also describes the effect of human predation on shellfish, fish, and birds as well as the impact of predators on humans, who were both competitors and potential prey.

Hemmer, Helmut. *Domestikation (Domestication: The Decline of Environmental Appreciation)*, trans. Niel Backhaus. New York: Press Syndicate of the University of Cambridge, 1990. ISBN 0 521 34178 7. 208 pp.

Hemmer provides an overview of the domestic animal and the process of domestication. The domestic animal is an essential element in the development of human civilization, and understanding domestication may help humans develop new kinds of domestic animals. These animals may improve food production for people throughout the world who live at a subsistence level, lead to alternative methods of land use, and lend themselves to intensive husbandry. New laboratory animals might help solve some of today's complex medical mysteries. A central issue is how the behavior of a domesticated animal differs from the behavior of its wild counterpart. The book discusses reactions to stress and psychological tolerance, behavioral flexibility, activity and intensity of action, aptitude for life in social groups, sexual and aggressive reactions, and pigmentation and body development in the domestication process.

Mason, Ian L., ed. *Evolution of Domesticated Animals.* New York: Longman, 1984. ISBN 0 582 46046 8.

This book presents an account of the origin and history of domestic animals, which are defined as animals whose breeding is under human control, that provide a product or service useful to humans, that are tame, and that have been selected away from the wild type (that is, a range of genotypes has been produced by selection). Mammals, birds, reptiles, amphibians, fish, insects, crustaceans, and mollusks are included.

Legal Issues: United States and International

Favre, David S. *International Trade in Endangered Species: A Guide to CITES.* Dordrecht, Netherlands: Martinus Nijhoff, 1989. ISBN 0 7923 0114 5. 418 pp.

The Convention on International Trade in Endangered Species of Wild Fauna and Flora, commonly called CITES, is an international treaty that attempts to oversee and control trade in wildlife. It went into force in 1975. The treaty consists of 17 articles; the first 10 set forth the substantive obligations of each state that is a party to the treaty. Most of the individual chapters in this book deal with an individual article of the treaty and provide the text of the article, a general discussion of the article, a historical note on the development of the language of the article, and selected notes.

————. *Wildlife, Cases, Laws and Policy.* Detroit, MI: Associated Faculty Press, 1983. ISBN 0 86733 023 6. 277 pp.

Favre provides an analysis of the key litigation dealing with the issues of state ownership of wildlife, the federal authority over migratory birds, horses and burros, and eagles, the National Environmental Policy Act, the Endangered Species Act, and the Marine Mammal Protection Act. He also discusses the importance of habitat.

————. *Wildlife Law.* Detroit, MI: Lupus Publications. 1991. ISBN 1 879581 03 5.

Favre provides an excellent overview and analysis of the litigation surrounding wildlife and endangered species. He discusses the state ownership doctrine, state authority and responsibility, the special role of Native Americans, the Endangered Species Act, legal rights for animals, the international trade in wildlife, and marine mammals.

Magel, Charles R. *Animal Rights.* Jefferson, NC: McFarland. 1989. ISBN 0 89950 405 1. 240 pp.

Magel presents an overview of the animal rights literature including an annotated bibliography arranged in chronological order, as well as a list of animal rights organizations.

Orloff, Neil. *The Environmental Impact Statement Process.* Washington, DC: Information Resources Press. 1978. ISBN 0 87815 021 8. 242 pp.

The National Environmental Policy Act mandates that before a federal department, agency, or bureau can undertake a project that might impact on the environment, it must produce an environmental impact statement. This is also required of projects by state and local governments and the private sector, if federal moneys are involved. Citizens can have input into what is covered by an impact statement and its completeness. The book provides the text of NEPA and the text of the regulations governing the development and use of environmental impact statements.

Reitze, Arnold W., Jr. *Environmental Planning: Law of Land and Resources.* Washington, DC: North American International, 1974. ISBN 0 88265 033 5.

This is a very useful book that provides an overview and interpretation of environmental laws and selected litigation. It covers endangered species, marine species, and forests and forest utilization, as well as wetlands and topics related to energy generation, mining, and pollution.

Rohlf, Daniel J. *The Endangered Species Act.* Stanford Environmental Law Society. 1989. 207 pp.

Rohlf provides a historical review of the endangered species legislation and an interpretation of some of its key components, especially Section 7 of the Endangered Species Act, which mandates that federal agencies must ensure that their activities do not jeopardize the continued existence of listed species or modify their habitat and Section 9, which lists the act prohibited by the Endangered Species Act.

Sherry, Clifford J. *Animal Rights: A Reference Handbook.* Santa Barbara, CA.: ABC-CLIO, 1994, ISBN 0 87436 733 6. 240 pp.

This book provides an unbiased overview of the controversies surrounding the issue of animal rights.

Sigler, William F. *Wildlife Law Enforcement.* Dubuque, IA: Wm. C. Brown. 1972. ISBN 0 697 08202 4. 360 pp.

Sigler provides a good summary of the pre–Endangered Species Act wildlife law and a list of important litigation. He provides a good introduction to the history of wildlife law, including the common law background of current law. He describes the state and federal laws and regulations, including separate chapters on the rights of private citizens, the rights of the individual, the search of motor vehicles, and an overview of the penalties for violating wildlife law. The role of the wildlife law enforcement officer is described in detail, including tactics of arrest, arresting dangerous criminals, and the rules of evidence.

Novels

Abbey, Edward *The Monkey Wrench Gang.* New York: Avon Books. 1975. ISBN 0 380 00741 X. 387p

Monkey wrenching is a name used by radical "eco-warriors" to describe their tactics. This novel describes the "monkey wrenching" activities of Dr. A. K. Sarvis, George W. Hayduke, Seldom Seen Smith, and Ms. B. Abbzug in the Four Corners area (near the borders of Arizona, New Mexico, Utah, and Colorado).

Quinn, Elizabeth. *Killer Whale.* New York: Pocket Books. 1997. ISBN 0 671 52770 3. 240 pp.

This is one of a series of mystery stories in which the lead character, Lauren Maxwell, is an investigator for Wild America Society.

Juvenile

A good many books have been written for young people about endangered species, extinction, and other issues. Most are well written. A small sampling are included here.

Silverstein, Alvin, Virginia and Robert Silverstein. *Saving Endangered Species.* Hillside, NJ: Enslow, 1993. ISBN 0 89490 402 7. 126 pp.

This book provides an overview of how and why animals come into danger of becoming extinct. It provides suggestions on how young people can help save endangered species, as well as lists of organizations that can provide additional information.

————. *The Spotted Owl.* Brookfield, CT: Millbrook Press, 1994. ISBN 1 56294 415 0. 63 pp.

The spotted owl has been at the center of a storm since it was first proposed to list it as an endangered species and the controversy continues to this day. This book highlights some of the sources of this controversy.

Sobol, Richard. *One More Elephant.* New York: Cobblehill Books, 1995. ISBN 0 525 65179 9. 31 pp.

This book describes what Uganda is doing to try to save its elephant population.

Stefoff, Rebecca. *Extinction.* New York: Chelsea House, 1992. ISBN 0 7910 1578 5. 127 pp.

This well-illustrated book describes how animals go from threatened to endangered to extinct and what we can do to potentially prevent this process.

Tesar, Jenny. *Endangered Habitats.* New York: Facts on File, 1992. ISBN 0 8160 2493 6. 112 pp.

This book describes what a habitat is, why it is important, how humans affect animal habitats, and what young people can do to help protect and enhance habitats.

Federal Publications

The federal government issues dozens of publications dealing with the Endangered Species Act and related laws. Most of these publications come from the Department of the Interior and the Department of Commerce.

American Forest Council, American Forest Resource Alliance, Associated Oregon Loggers, California Forestry Association, Douglas Timber Operators, National Forest Products Association, Northwest Forest Resource Council, Northwest Forestry Association, Northwest Independent Forest Manufacturers Association, Oregon Forest Industries Council, Southern Oregon Timber Industries Association, Washington Citizens for World Trade, Washington Contract Loggers Association, Washington Forest Protection Association, Western Forest Industries Association, and Western Wood Products Association. *Comments on the Recovery Plan for the Northern Spotted Owl—Draft,* April 1992.

These organizations provide a detailed analysis of the Recover Plan, including their own studies and mathematical models. Their studies and models lead to different conclusions than found in the Recovery Plan.

Comments on the Designation of Critical Habitat for the Northern Spotted Owl, Proposed by the U. S. Fish and Wildlife Service, May 6, 1991 (26 Fed. Reg. 20816). Report to the Congress by the Comptroller General of the United States.

These organizations provide a detailed analysis of the critical habitat and find flaws in the reasoning.

Endangered Species—A Controversial Issue Needing Resolution. (GAI.13: CED 79–65) U.S. General Accounting Office. July 2, 1979.

The document provides a detailed look at the legislative history of the endangered species program and an overview of the responsibilities of organizations with the federal government.

Lujan, Manuel, Jr., Donald R. Knowles, John Turner, Marvin Plenert, and Jonathan Bart. *Recovery Plan for the Northern Spotted Owl—Draft*. April 1992. 662 pp.

This book provides a relatively comprehensive review of the biology of the northern spotted owl and its status under the Endangered Species Act. It also provides a detailed analysis of its current status, the perceived threats to this status, current management practices, and a very detailed recovery plan. Considering the controversy surrounding this owl, its management, and the amount of litigation that has occurred, it is interesting what is not analyzed in detail. "There are no estimates of the historical population size of the northern spotted owls, but the owls are *believed* to have inhabited most old-growth forests throughout the Pacific Northwest and northwestern California, . . . (emphasis added)."

Lujan, Manuel, Jr., Donald R. Knowles, John Turner, Marvin Plenert, and Jonathan Bart. *Summary of the Draft Recovery Plan for the Northern Spotted Owl*. April 1992.

This book is a summary of the recovery plan for the northern spotted owl (see above).

United States Department of Agriculture. *Fact Sheet: Animal Damage Control.* July 1991.

———. *Fact Sheet: Animal Damage Control.* August 1991.

———. *Fact Sheet: Animal Damage Control.* March 1992.

United States Fish and Wildlife Service. *Endangered Means There's Still Time.* 1981.

———. *Endangered Species.* (I 49.2 EN 2/23/993) 1993.

This brochure provides an overview of the controversies surrounding endangered species and the Endangered Species Act.

———. *Endangered Species: The Road to Recovery* (I 49.2 EN 2/5).

This brochure highlights some of the successes of the Endangered Species Act, such as the recovery of the whooping crane. It describes how recovery programs come about and the role of the recovery team, habitat acquisition, manipulation, and cleanup, as well the transplantation of endangered species to new habitat and the role of captive breeding.

———. *Placing Animals and Plants on the List of Endangered and Threatened Species.* (I 49.2 EN 2/10/993) 1993.

This brochure described the process used in listing a species under the Endangered Species Act and how critical habitat is defined.

———. *Why Save Endangered Species?* (I 49.2 EN 2/7/993) 1993.

This brochure describes why biodiversity is important and when it is important to save endangered species.

Law School Reviews

Law school reviews are edited by law school students chosen on the basis of their scholarship record or through a writing competition. The typical review is divided into several sections. One section, commonly called comments, is generally written by law professors and are scholarly in nature. These comments often have a significant impact on changing the law or charting the course of new developing fields of the law. The law students write notes that deal with surveys of selected topics or critical analyses of current court cases. The courts often cite law review articles and student commentaries

Blumm, Michael C. **"Ancient Forests, Spotted Owls, and Modern Public Land Law."** *Boston College Environmental Affairs Law Review* 18 (1990): 605–622.

Bonnett, Mark, and Kurt Zimmerman. **"Politics and Preservation: The Endangered Species Act and the Northern Spotted Owl."** *Ecology Law Quarterly* 18 (1991): 105–171.

Foley, Elizabeth A. **"The Tarnishing of an Environmental Jewel: The Endangered Species Act and the Northern Spotted Owl."** *Land Use and Environmental Law* 8 (1992): 253–283.

Goplerud, C. Peter. **"The Endangered Species Act: Does It Jeopardize the Continued Existence of Species?"** *Arizona State Law Journal* (1979): 487–510.

Rosenberg, Ronald H. **"Federal Protection of Unique Environmental Interests: Endangered and Threatened Species."** *North Carolina Law Review* 58 (1980): 491–559.

Sher, Victor M., and Carol Sue Hunting. **"Eroding the Landscape, Eroding the Laws: Congressional Exemptions from Judicial Review of Environmental Laws."** *Harvard Environmental Law Review* 15 (1991): 435–491.

Simmons, Robert M. **"The Endangered Species Act of 1973."** *South Dakota Law Review* 23 (1978): 302–325.

Nonprint Resources 8

M any nonprint sources offer information about the controversy surrounding endangered species. These include on-line and CD-ROM databases (sometimes also available in print), multimedia presentations, the Internet, including the World Wide Web, and audio- and videotapes. This chapter provides an overview of these resources.

Databases

Access
Provider: John Gordon Burke
              Publishing, Evanston, IL
Coverage: 1988–
Updates: Semiannually
Data Type: Bibliographic with abstracts

Access is a periodical index that complements and supplements the *Reader's Guide to Periodical Literature,* providing information about magazines not covered by the *Reader's Guide.* Access contains more than 2 million records, with about 100,000 added each year.

Agricola
Provider:	U.S. National Agricultural Library, Beltsville, MD
Coverage:	1970–
Updates:	Monthly
Data Type:	Bibliographic

Agricola, which contains more than 2 million records, indexes U.S. federal agencies' publications, United Nations Food and Agriculture Organization (FAO) reports, publications of state agricultural experiment stations and extension services, and approximately 600 periodicals. The database provides worldwide coverage of the literature of agricultural science and related topics, including the welfare of animals used in exhibition, education, and research.

Analytical Abstracts
Provider:	Royal Society of Chemistry, Cambridge, UK
Coverage:	1980–
Updates:	Monthly
Data Type:	Bibliographic

Analytical Abstracts references more than 1,300 journals that deal with all aspects of analytical chemistry, including environmental chemistry.

Biological Abstracts
Provider:	Biosis, Philadelphia
Coverage:	1969–
Updates:	Weekly
Data Type:	Bibliographic with abstracts

Biological Abstracts provides coverage of the worldwide literature on life science. Approximately 6,000 international journals are monitored on all aspects of life science, including ecology.

Biological Abstracts (RRM)
Provider: Biosis, Philadelphia
Coverage: 1989–
Updates: Quarterly
Data Type: Bibliographic

Biological Abstracts (RRM) provides worldwide coverage of life-science meetings, books, and reviews, as well as U.S. patent information. More than 2,000 meetings and symposia are covered.

CA Search
Provider:	Chemical Abstracts Service, Columbus, OH

Coverage: 1967–
Updates: Biweekly
Data Type: Bibliographic

CA Search contains more than 10 million citations to the worldwide literature on chemistry and its applications. The database can be searched by CAS registry numbers, each of which is a unique number assigned to a chemical compound.

Commonwealth Agricultural Bureaux (CAB)
Provider: CAB International, Slough, United Kingdom
Coverage: 1972–
Updates: Weekly
Data Type: Bibliographic

CAB Access covers more than 1,000 journals worldwide that contain information about agriculture, forestry, and related sciences. About 1,800 items are added each week.

Conference Papers Index
Provider: Cambridge Scientific Abstracts, Bethesda, MD
Coverage: 1973–
Updates: Bimonthly
Data Type: Bibliographic

Conference Papers Index provides access to papers presented at more than 1,000 major regional, national, and international scientific and technical meetings each year.

CQ Researcher
Provider: Congressional Quarterly, Washington, DC
Coverage: 1992–
Updates: Quarterly
Data Type: Full text

CQ Researcher provides in-depth, nonpartisan summaries on the hottest issues of the day and provides leads to factual and objective information about controversial subjects.

Current Contents Search
Provider: Institute for Scientific Information, Philadelphia
Coverage: Current
Updates: Weekly
Data Type: Bibliographic

Current Contents Search allows searches of the table of contents pages of leading journals in the sciences. Most databases have a time lag between when an article appears in a journal (or a presentation at a meeting yields an abstract) and when the citation or abstract appears in the databases, even though the lag time has been decreasing over the last decade. In contrast, Current Contents Search provides virtually immediate access to articles because it receives contents pages at the time journals are published, and sometimes well before.

Dissertation Abstracts Online
Provider:　University Microfilms International,
　　　　　　 Ann Arbor, MI
Coverage:　1861–
Updates:　 Monthly
Data Type:　Bibliographic

Dissertation Abstracts Online provides an author, title, and subject guide to virtually every doctoral dissertation accepted by accredited academic institutions since 1861, when the first academic doctoral degree was granted.

Ecodisc
Provider:　Elsevier Science
Coverage:　1990–
Updates:　 Quarterly
Data Type:　Bibliographic with abstracts

Ecodisc contains 70,000 citations to the worldwide literature on ecology and the ecosystem. It includes 2,000 primary journals, as well as books, monographs, and reports.

EIS
Provider:　Cambridge Scientific Abstracts, Bethesda, MD
Coverage:　1985–
Updates:　 Quarterly
Data Type:　Bibliographic

EIS contains abstracts of all the environmental impact statements issued by the federal government since 1985.

Encyclopedia of Associations
Provider:　Gale Research, Detroit, MI
Coverage:　Current

Updates: Quarterly
Data Type: Descriptions

Encyclopedia of Associations provides up-to-date information about international, national, regional, state, and local associations.

Environmental Quality
Provider: CAB International, Slough, UK
Coverage: 1973–
Updates: Quarterly
Data Type: Bibliographic

Environmental Quality covers environmental issues such as soil and erosion, the impact of tourism on the environment, deforestation, and forest decline.

Federal Research in Progress
Provider: National Technical Information Service, Springfield, VA
Coverage: Current
Updates: Monthly
Data Type: Bibliographic

Federal Research in Progress provides access to information about federally funded research projects. Each record contains the title of the project, the name of the principal investigator, the names of the organizations performing and sponsoring the work, and a description of the research.

Gale Directory of Databases
Provider: Gale Research
Coverage: Current
Updates: Semiannually
Data Type: Bibliographic

The Gale Directory of Databases provides worldwide information about more than 5,300 databases available on-line as well as 3,500 databases on CD-ROM, floppy disk, and magnetic tape.

Geobase
Provider: Elsevier Science
Coverage: 1980–
Updates: Quarterly
Data Type: Bibliographic

Geobase provides worldwide coverage of more than 2, 000 primary journals, books, monographs, and reports on ecology and human geography, including environmental and Green issues.

Life Sciences Collection
Provider: Cambridge Scientific Abstracts, Bethesda, MD
Coverage: 1978–
Updates: Monthly
Data Type: Bibliographic citations with abstracts

The Life Sciences Collection contains abstracts of the literature of the life sciences. The collection covers the fields of animal behavior, biochemistry, ecology, entomology, genetics, immunology, microbiology, oncology, neuroscience, toxicology, and virology.

Magazine Index
Provider: Information Access Co., Foster City, CA
Coverage: 1973–
Updates: Weekly
Data Type: Bibliographic

Magazine Index provides coverage of more than 500 popular magazines.

National Newspaper Index
Provider: Information Access Co., Foster City, CA
Coverage: 1982–
Updates: Weekly
Data Type: Bibliographic

National Newspaper Index indexes the *New York Times,* the *Wall Street Journal,* the *Christian Science Monitor,* the *Washington Post,* and the *Los Angeles Times.*

Newsearch
Provider: Information Access Co., Foster City, CA
Coverage: Current month only
Updates: Daily
Data Type: Bibliographic

Newsearch provides a daily index of news stories and articles from 1,700 newspapers, magazines, and periodicals.

NTIS
Provider: National Technical Information Service, Springfield, VA

Coverage: 1964–
Updates: Biweekly
Data Type: Bibliographic

The NTIS provides access to the results of government-sponsored research, development, engineering, and analysis.

OCLC (On-Line Computer Library Center)
Environment Library
Provider: OCLC, Dublin, OH
Coverage: 1980–
Updates: Annually
Data Type: Bibliographic

The OCLC Environment Library provides more than 670,000 records, drawn from the OCLC Union Catalog, that deal with the environment and environmental issues.

Periodical Abstracts Research
Provider: UMI, Ann Arbor, MI
Coverage: 1986–
Updates: Monthly
Data Type: Bibliographic with abstracts and with full text
                 from 1992 on

Periodical Abstracts Research provides citations to more than 1,600 general-interest periodicals.

Reader's Guide to Periodical Literature
Provider: H. W. Wilson Co.
Coverage: 1983–
Updates: Monthly
Data Type: Bibliographic

The Reader's Guide to Periodical Literature provides coverage of 240 popular magazines.

Scisearch
Provider: Institute for Scientific Information, Philadelphia
Coverage: 1974–
Updates: Weekly
Data Type: Bibliographic

Scisearch enables the retrieval of newly published articles based on an author's reference to prior articles. For example, if a search of Medline and Agricola yields information about an article written several years ago that described a particular in vitro test, then Scisearch can be searched for the author's name. The Scisearch search yields a list of all papers that cite that author's work, including complete bibliographic references. Scisearch enables an investigator to assess the impact of an idea or a technique. If no one cites a particular paper, it may mean the article had minimal or no impact. If many people cite the paper, it means that the ideas or techniques it describes have had an impact. Scisearch also allows searches by text words via the Permutext index. This is especially valuable for new terms, because they often appear as Permutext entries before they become available as text terms in other databases.

TreeCD
Provider: CAB International, Slough, UK
Coverage: 1939–
Updates: Quarterly
Data Type: Bibliographic

TreeCD contains more than 300,000 abstracts covering the national and international forestry literature.

Zoological Record on CD
Provider: Biosis, Philadelphia
Coverage: 1978–
Updates: Quarterly
Data Type: Bibliographic

Zoological Record on CD provides worldwide coverage of the zoological literature.

The Internet

One important thing to remember about the Internet is that anyone with a computer, a modem, a telephone, and a little technical knowledge can create a Web page and put virtually anything on it. Essentially nothing on the Web comes under editorial scrutiny, so it is important to take what you find there with a grain of salt.

Laws and Legal Issues

Thomas: Legislative Information on the Internet
Internet Address: http://thomas.loc.gov

Provides access to information about congressional activity, including bills before the House of Representative and the Senate, tracked by topic, popular or short name, and bill number. It is possible to track current activity on the reauthorization of the Endangered Species Act, as well as appropriations for environmental activities.

U.S. House of Representatives Internet Law Library U.S. Code
Internet Address: http://law.house.gov/usc.htm

The U.S. Code is the official compilation, in subject matter order, of federal law of a general and permanent nature that is currently in force. The code is divided into 50 titles, and each title is divided into sections. This on-line version of the code allows you to search by words or concepts.

U.S. House of Representatives Internet Law Library Code of Federal Regulations
Internet Address: http://www.access.gpo.gov/nara/cfr/crf-table-search.hml

The *U.S. Code of Federal Regulations* is the official compilation, in subject matter order, of federal regulations of general applicability and legal effect that are currently in force. This on-line version allows you to search the code by keyword, phrase, topic, title, and part number. For example, Title 50, Chapter IV, Subchapter A, Part 424.12 sets the criteria for designating critical habitat for an endangered species.

Directories

The Amazing Environmental Organization Web Directory
Internet Address: http://www.webdirectory.com

This directory has collected and annotated Web pages dealing with environmental issues and ecology, including endangered species.

General

Biodiversity and Biological Collections Web Server
Internet Address: http://muse.bio.cornell.edu

This page provides an overview of biodiversity and why it is important. The site has links to other pages on all levels of the phylogenetic tree.

Conservation Hypertext Book
Internet Address: http://www.bio.uci.edu/academic/ungrad/resource.html

This is a hypertext textbook on the history of life on Earth. It provides discussion and links to other sites on endangered and extinct species. For example, in the early nineteenth century, billions of passenger pigeons *(Ectopistes migratorius)* lived in vast flocks in the forests of the eastern United States. Martha, the last surviving passenger pigeon, died at 1:00 P.M. on 1 September 1914 in the Cincinnati Zoological Gardens. You can find an image of Martha at http://straylight.tamu.edu/bene/lg/losses.html.

Discovery Channel Online
Internet Address: http://www.discovery.com/DCO/doc/1012/tools.search/search.html

This is an on-line service of the Discovery Channel that allows searches to be conducted in plain English, using a keyword or phrase.

EcoNet
Internet Address: http://www.econet.apc.org/econet/

This site provides access to current news and information for and about people interested in environmental preservation and sustainability.

The Ecology Channel
Internet Address: http://www.ecology.com

This site provides information about ecology and the impact of humans on the environment.

EcoNet: Endangered Species
Internet Address: http://www.econet.apc.org/endangered/

This page provides information on current legislation as well as position papers about endangered species and the Endangered Species Act.

EE-Link Endangered Species
Internet Address: http://www.nceet.snre.umich.edu/EndSpp/Endangered.htm/

This site provides lists of threatened and endangered species, sorted by region and group. It also provides images of selected endangered species.

EnviroLink/Endangered Species
Internet Address: http://envirolink.org/elib/issues/esa/

This page provides an overview of the Endangered Species Act of 1973 and its amendments, as well as interpretation and discussions of reauthorization and implementation.

The Greens
Internet Address: http://www.dru.nl/maatschappij/politiek/groenen/intlhome.htm; *North America:* http://www/greems/org.

The Greens constitute a worldwide activist and political movement that promotes ecological wisdom, social justice, grassroots democracy, and nonviolence.

International Institute for Sustainable Development
Internet Address: http://www.iisd.ca/linkages/index.html

This page provides information and updates on the Convention on Biological Diversity and on other issues related to threatened and endangered species.

News Link Environmental
Internet Address: http://www.caprep.com/caprep/

This page provides access to current environmental news as well as to state and federal legislation and regulations affecting the environment.

U.S. Fish and Wildlife Service Endangered Species Program
Internet Address: http://www.fws.gov/~r9endspp/endspp.html

The U.S. Fish and Wildlife Service is one of the agencies charged with administering the Endangered Species Act. Its page on endangered species provides on-line access to the Endangered Species Bulletin as well as the full text of the Endangered Species Act of 1973.

Virtual Library Environment and Ecology Sections
Internet Address: http://www.infi.net/~cwt/env.html

This site provides information about ecology and the environment, with lots of links to other pages of interest.

Wired for Conservation
Internet Address: http//www.tnc.org

This is the Web page of the Nature Conservancy, which uses private funds to buy land and water needed to preserve endangered habitats and species. The organization manages more than 1,500 private nature sanctuaries in the United States.

CD-ROMs

Encyclopedia of U.S. Endangered Species

This is a CD-ROM published in 1994 by Zane Publishing (Dallas, TX) that contains multimedia presentations about the endangered species of plants and animals of the United States. The CD contains 836 reports about threatened and endangered species. Each report contains a photo of the species, a map showing the location where the species is found, a pronunciation guide for both the common and scientific name, the status of the species (threatened, endangered), its taxonomy, a description of the species, its habitat, food, reproductive status, and why it is threatened. Each report contains hypertext (highlighted) words. If you click on one of these words, a glossary opens that defines the word. The main menu contains submenus that allow the reports to be located by searching by the common or scientific name of the species, by the state where it is located, or by its taxonomic group (for example, birds, fish).

Animals of the World

This CD contains descriptive text, photographs, video, and audio about common and exotic animals, from aardvarks to zebras.

The Big Green Disc

Problems facing Earth, such as global warming, pollution, ozone depletion, and acid rain, are highlighted on this CD. Leading experts discuss these problems and how to overcome them.

Biomes, Volumes I and II

These CDs offer text and audiovisual presentations that describe the distribution, climates, topography, and general appearance of biomes as well as the unique features and adaptations of the plants and animals that inhabit each biome. Volume I presents the deciduous and coniferous forests of North America and the tropical rain forest and savanna biomes. Volume II examines the grassland, tundra, desert, and chaparral biomes.

Discovering Endangered Wildlife

Video, photographs, and text describe how animals live, their habits, what they eat, and the threats they face.

Ecodisc

This interactive CD makes you the manager of a wildlife reserve and lets you see how your actions affect the plants and animals in the reserve. It also allows you to see the impact of special interest groups, such as fishers, hunters, and farmers.

Endangered and Threatened Species

The U.S. Fish and Wildlife Service provides information about plants, animals, birds, amphibians, fish, and invertebrates that are on the worldwide endangered species list. This CD also highlights wildlife recovery programs and what programs are being used to increase the population of these species.

Environmental Science: Field Library

This interactive CD allows you to experience the collection and evaluation of environmental data about topics such as stream pollution, stream flooding, and legal control of the environment.

Explorapedia: The World of Nature

This CD, designed for young people, provides an introduction to rain forests, lakes, deserts, grasslands, mountains, savannas, deciduous forests, and coral reefs.

Five Kingdoms: Life on Earth

This interactive CD provides multimedia and hypertext information about the 92 phyla living today, plus selected genera.

Grzimek's Multimedia Encyclopedia of Mammalian Biology

This CD provides a visual description of the world's ecosystems and an encyclopedic coverage of mammalian biology.

Last Chance to See

This CD by Douglas Adams *(The Hitchhiker's Guide to the Galaxy)* contains text and photographs that describe endangered species from Africa, China, and the South Pacific. It also contains the text of Douglas's book *The Last Chance to See*.

Computer Simulation Programs/ Virtual Reality

Ecology is the study of the interrelationships of organisms and the physical environment. An ecosystem is a community of plants and animals in an environment that supplies the raw materials for life. When an ecosystem is disrupted, one or more species may be threatened by extinction. These disruptions can be either natural, such as drought, or man-made, such as overexploitation.

One way to begin to understand the impact of change in an ecosystem is to use a computer model or simulation. Prior to the 1980s, most simulations were made up of line after line of computer code and equations. The output of the simulations was graphs or simple line drawings. The development of powerful desktop computers and high-density storage devices (for example, interactive CD-ROMs) has allowed simulation designers and model builders to add three-dimensional graphics as well as photographs, audio, and video to simulations. The people who use these simulations range from laypeople interested in ecology and endangered species to high school and college students to graduate students and professionals.

Although still in its infancy, virtual reality holds much promise. Virtual reality technology (hardware and software) allows you to step into a computer-generated world—for example, an ecosystem—where you can look around, move around, and interact with the environment. The ecosystem can be one that really exists, one that existed in the past (for example, when dinosaurs roamed Earth), or one that could exist in the future (for example, when global warming has taken place). Special goggles allow you to see this artificial world. Earphones or speakers provide sound. Tactile and force sensations are more difficult to model; it is unlikely that the same level of realism will be achieved with these sensory systems. Tracking devices, such as wired gloves and clothing and motion detectors, monitor body parts. Interaction devices enable the user to manipulate virtual objects. It is likely that virtual reality technology will continue to mature and its cost decrease. So computer simulations and virtual reality promise to provide more and more powerful tools to study endangered species as well as ecology, population genetics, and related topics.

It is important to note that to develop a model or a simulation, the modeler must understand the topic he or she is attempting to model. Although amazing strides have been made in ecology in the last few decades, especially with regard to endangered species, it is unlikely that computer simulations will replace basic experimental research in this area because many basic facts are still not known about ecosystems and human impact on them.

SimEarth: The Living Planet (Maxis, Walnut Creek, CA) is an example of a relatively user-friendly simulation. It is based on the controversial Gaia theory of James Lovelock. Lovelock believes that we must view the entire Earth as a living planet—not just the biosphere, not just the living organisms, but the whole planet, including the molten core, the rocks, the air, and the oceans. He believes that the planet is a single living, evolving entity. SimEarth provides you with a set of rules and tools that describe, create, and control an Earthlike planet. A complicated system, such as a planet, cannot be completely defined by a series of formulas and equations, so SimEarth, like all simulations, is an abstraction. It allows you to fix the parameters of the geosphere, the hydrosphere, the atmosphere, and the biosphere and determine the impact of each of your decisions. You can introduce natural catastrophes, such as volcanoes, earthquakes, meteors, and tidal waves, and determine their impact. You can also simulate a Marslike or a Venuslike planet.

Selected Models/Simulations

EcoBeaker
Internet Address: http://www.webcom.com/sinauer/ecobeaker.html

This simulation allows you to place creatures whose behavior you design into a two-dimensional world where they eat, reproduce, move around, and die. You can create as many as ten species in up to ten habitats. The laboratory guide explains each model and how you can construct your own model.

Fish Farm
Internet Address: http://129.78.177.135/RO.eld+HTMLRecord& RecID=513

This simulation allows you to set up a fish farm and determine the best environmental and feeding conditions for different species of fish.

Wa-Tor for Windows
Internet Address: http://nhsbig.inhs . . . ducational/WARORW. README

This simulation enables you to set up a world of fish and sharks and allows you to determine how often the fish and sharks breed and eat.

Ecological Simulations

You can find more information about ecological simulations and models at a number of sites on the World Wide Web. Some of the home pages contain executable programs that can be down-loaded to your computer, whereas others provide information about programs that you can purchase. Some of these simula-tions are meant for students, while others are designed for the knowledgeable professional.

Biology Education Software FAQ
http://www.zoology.washington.edu.biosoft/

The BioQuest Library
http://terrapin.umd.edu/bioquest.html

CTI Center for Biology
http://www.liv.ac.uk/ctibiol.html

Illinois Natural History Survey Wildlife Ecology Software Server
http://www.inhs.uius.edu/ftpsvr.cweftp/ftphome.html

Ramas: Ecological and Environmental Software
http://gramercy.ios.com/~ramas/

UniServe Science
http://www.usyd.edu.au/su/SCH/

Selected Audio- and Videotapes

The America's View of Animals
Source: University of Minnesota, Minneapolis
Format: ½" VHS videotape
Length: 45 minutes

This tape presents a discussion of how current attitudes toward animals and human-animal relationships are influenced by religious beliefs (Jewish, Catholic, Protestant, and Native American) and historical conventions.

At the Crossroads
Source: Stroffer Productions, Penn State
Format: Film
Length: 26 minutes

This film provides an overview of the dangers faced by North American wildlife.

Animal Behavior in the Wild
Source: University of Minnesota, Minneapolis
Format: ½" VHS videotape
Length: 45 minutes

Richard Phillips and David Smith of the University of Minnesota illustrate aspects of animal behavior and communication.

Animals in Zoos: Issues and Concerns
Source: University of Minnesota, Minneapolis
Format: ½" VHS videotape
Length: 45 minutes

A panel of experts discusses the role of zoos in maintaining genetic diversity in captive animal populations.

The Business of Extinction
Source: King Features Entertainment, WGBH
Format: Film
Length: 50 minutes

Originally shown as part of the *Nova* educational series, this film highlights the illegal trade in endangered species. The effects of collecting wild animals for pets and zoos are shown, as is the role of the Convention on International Trade in Endangered Species.

Exodus at Yellowstone: The Second Catastrophe
Source: Animal Protection Institute of America, Sacramento, CA
Format: ½" VHS videotape
Length: 30 minutes

Animal protection advocates voice their protests over bison hunting, focusing on the bison that were shot after straying outside Yellowstone National Park boundaries in the winter of 1989, following a severe drought and a fire that destroyed food supplies within the park.

The Galapagos: Darwin's World within Itself
Source: Encyclopedia Britannica Educational Corp.
Format: Film
Length: 20 minutes

This film shows the Galápagos Islands' unique environment and wildlife, some of which does not exist anywhere else in the world. The movie also shows the human impact on this fragile ecosystem.

The Image of Animals Today
Source: University of Minnesota, Minneapolis
Format: ½" VHS videotape
Length: 45 minutes

A panel discusses the portrayal of animals in contemporary American mass media, including cartoons, books, poems, and television. The panel formulates a general conception of America's image of animals.

Last Stronghold of the Eagle
Source: National Audubon Society
Format: Film
Length: 30 minutes

This film provides one of the best introductions to the bald eagle, our national symbol. Its vocalizations, its relations with other bald eagles, its feeding, and its flight are covered in this beautifully made film

Memories from Eden
Source: Time Life Films, WGBH
Format: Film
Length: 57 minutes

This *Nova* film shows the role of zoos as educators and conservators of endangered wildlife.

Vanishing Prairie: Buffalo—Majestic Symbol of the American Plains
Source: Disney
Format: Film
Length: 12 minutes

This film is a bit dated but is in the best tradition of the Disney nature films, which means that the photography is superb.

Glossary

CFR Code of Federal Regulations

CITES Convention on International Trade in Endangered Species of Wild Fauna and Flora, adopted in 1973

convention An international agreement made prior to a formal treaty or agreement for the regulation of matters of mutual concern

critical habitat All the physical and biological features that are essential to the conservation of a given species

endangered species Any species that is in danger of extinction throughout all or a significant portion of its range

Endangered Species Act The Endangered Species Act (ESA) of 1973 and its amendments, which require the federal government to list species that are threatened and endangered in the United States and prohibits hunting and destruction of critical habitst of these species

IUCN International Union for the Protection of Nature/World Conservation Union, which publishes the Red List of Threatened Species

natural selection The process whereby variability from individual to individual allows some individuals to survive and prosper in a given environment, while others perish

recovery plan A formal statement of what needs to be done to reestablish a species so that it can be removed from the list of threatened or endangered species

species   A group of successful organisms that share an environmental niche, tend to have similar morphological and behavioral characteristics, can mate and produce viable offspring, and share a common gene pool.

speciation   The development of new subspecies or species, which is most likely to occur when environments are changing rapidly, when chromosomal mutations occur, or when a population becomes geographically separated from the rest of its species

Threatened Species   Any species that is likely to become endangered within the foreseeable future throughout all or a significant portion of its range.

treaty   A formal compact made between two or more independent nations about matters of mutual concern

USC   The official compilation of U.S. statutes arranged according to topic

Appendix: Geological Timeline

EON/ERA/*Period*/Epoch	Millions of Years Ago	Life Forms	Events
PHANEROZOIC			
CENOZOIC			
Quaternary			
Holocene	0–0.1	Humans	
Pleistocene	0.1–2	Human ancestors	Ice ages
Tertiary-Neogene			
Pliocene	2–5		
Miocene	5–24	Modern mammals, horses, dogs, primates	
Tertiary-Paleogene			
Oligocene	24–37	Rise of modern mammals	
Eocene	37–58	Mammals abundant, rodents, carnivores, whales	
Paleocene	58–66		
MESOZOIC			
Cretaceous	66–144	Flowering plants	Cretaceous-Tertiary (K-T) extinction, all 65 MYA, 85% of all species disappeared, including dinosaurs, minimal effect on mammals, birds, small reptiles, and amphibians
Jurassic	144–208	Birds/mammals emerge	
Triassic	208–245	Dinosaurs	
PALEOZOIC			
Permian	245–286		90–95% of all marine species eliminated
Carboniferous Pennsylvanian	286–320	Reptiles	
Carboniferous Mississippian	320–360	Trees	

EON/ERA/*Period*/Epoch	Millions of Years Ago	Life Forms	Events
Devonian	360–408	Amphibians	70% of the marine taxa perished, especially reef-building organisms
Silurian	408–438	Land plants	
Ordovician	438–505	Fish	440–450 MYA, 1/3 of all brachiopods and bryozoan families disappeared
Cambrian	505–570	Shells, Trilobites	Oldest group of trilobites and archaeocyathids (reef-building organisms) went extinct. Three other major extinctions occurred
PROTEROZOIC	570–2500	Multicelled organisms	650 MYA, 70% of the flora and fauna perished
ARCHEAN	2500–3800	Single-celled organisms	
HADEAN	3800–4600		

Sources: Carol O. Dunbar, *Historical Geology* (New York: John Wiley, 1949); William K. Hartman and Ron Miller, *The History of Earth* (New York: Workman, 1991).

Index

259

262 Index

268 Index

Clifford J. Sherry, Ph.D., is a senior scientist and principal investigator with Veridian. He is the author of six other books, including *Animal Rights* (ABC-CLIO), *Mathematics of Technical Analysis,* and *Drugs and Eating Disorders.* His professional interests include neurobiology, psychopharmacology, reproductive physiology/behavior, and teratology.